Parag Sewalkar

A Framework for Quality of Service in Vehicle-to-Pedestrian Safety Communication

A Framework for Quality of Service in Vehicle-to-Pedestrian Safety Communication

Parag Sewalkar

Universitätsverlag Ilmenau
2021

Impressum

Bibliografische Information der Deutschen Nationalbibliothek
Die Deutsche Nationalbibliothek verzeichnet diese Publikation in der Deutschen Nationalbibliografie; detaillierte bibliografische Angaben sind im Internet über http://dnb.d-nb.de abrufbar.

Diese Arbeit hat der Fakultät für Elektrotechnik und Informationstechnik der Technischen Universität Ilmenau als Dissertation vorgelegen.

Tag der Einreichung:	4. Mai 2020
1. Gutachter:	Univ.-Prof. Dr. rer. nat. Jochen Seitz (Technische Universität Ilmenau)
2. Gutachter:	Univ.-Prof. Dr.-Ing. Günter Schäfer (Technische Universität Ilmenau)
3. Gutachter:	Univ.-Prof. Dr.-Ing habil. Falko Dressler (Technische Universität Berlin)
Tag der Verteidigung:	12. Januar 2021

Technische Universität Ilmenau/Universitätsbibliothek
Universitätsverlag Ilmenau
Postfach 10 05 65
98684 Ilmenau
https://www.tu-ilmenau.de/universitaetsverlag

ISBN 978-3-86360-239-0 (Druckausgabe)
DOI 10.22032/dbt.47972
URN urn:nbn:de:gbv:ilm1-2021000031

Dedicated to my grandparents, Late Shree. Raghuraj S. Joshi and Smt. Minaxi R. Joshi

Acknowledgments

I dream things that never were and say "Why Not"?

- George Bernard Shaw

This quote has inspired me to explore many avenues including the decision to embark on a PhD journey.

First and foremost, I would like to thank Prof. Jochen Seitz for his unconditional and relentless support during my PhD. As I explored various paths during my PhD, from part-time research with another organization to the formation of our own consortium to my latest startup venture, it could not have been possible without his support. I would like to thank the reviewers of my dissertation and also, committee members. Their feedback and insights have been really helpful.

I would like to thank my Communication Networks teammates Maik, Abdulla, Matthias, and Mr. Heubach for keeping the team spirit vibrant. I thank Ms. Patotschka for supporting me in numerous administrative processes. I would also like to thank Gabriel and Mendrit for their immense support during the last phase of my PhD.

Finally, I thank my family and friends for their continued support. They've always inspired me to better myself, directly or indirectly.

Abstract

Vehicle-to-Everything (V2X) communication has emerged as an important mechanism to improve the safety and efficiency of road traffic. V2X communication encompasses Vehicle-to-Vehicle (V2V), Vehicle-to-Infrastructure (V2I), and Vehicle-to-Pedestrian (V2P) communication. Among these types, the V2P communication efforts continue to be in the preliminary stage and lack a rounded approach towards the development of V2P systems. V2P involves communication between vehicles and a wide variety of Vulnerable Road Users (VRUs), such as pedestrians, bicyclists, mopeds, etc. The V2X systems were originally developed only for V2V and V2I when solely the vehicle characteristics were in focus. However, effective V2P system design needs to consider the characteristics of VRUs. The differing characteristics of VRUs have given rise to many questions while adapting to the V2V communication model for the V2P system. This dissertation addresses three aspects pertaining to the development of the V2P safety system. The first aspect involves a systematic design of a V2P system using a holistic approach. This dissertation proposes a V2P design framework based on various categories of inputs that are required for the design of an effective V2P system. This framework improves the understanding of the V2P system requirements and helps make the design process more systematic. The second aspect is the network performance of the V2X network in the presence of a large number of VRUs. This dissertation proposes MC-COCO4V2P, which is an energy-efficient pedestrian clustering mechanism for network congestion mitigation. MC-COCO4V2P improves network performance by reducing the pedestrian-generated safety messages. It also improves the battery life of the pedestrian devices in the process. The third aspect involves the reliability of communication between a pair of a vehicle and a pedestrian that are on the verge of collision. This dissertation classifies such crucial communication as the one requiring the highest priority even among the exchange of critical safety messages. It proposes a mechanism enabling the surrounding nodes to reduce the communication priority temporarily. This results in preferred medium access for the pair resulting in higher Quality-of-Service (QoS) for the crucial communication.

Kurzfassung

Kommunikation zwischen Verkehrsteilnehmern (V2X) hat sich zu einem wichtigen Mechanismus zur Verbesserung der Sicherheit und Effizienz des Straßenverkehrs entwickelt. Obwohl die V2X-Kommunikation prinzipiell die Kommunikation zwischen Fahrzeugen (V2V), zwischen Fahrzeug und Infrastruktur (V2I) sowie zwischen Fahrzeug und Fußgänger (V2P) umfasst, sind Ansätze zur V2P-Kommunikation weiterhin in einem sehr frühen Stadium und lassen einen umfassenden Ansatz für die Entwicklung von V2P-Systemen vermissen. V2P umfasst im Detail die Kommunikation zwischen Fahrzeugen und einer Vielzahl von gefährdeten Verkehrsteilnehmern (VRUs), wie beispielsweise Fußgänger, Radfahrer oder Mopeds. V2X-Systeme wurden ursprünglich nur für V2V- und V2I-Kommunikation entwickelt, wobei ausschließlich die Fahrzeugeigenschaften im Fokus standen. Ein effektives V2P-Systemdesign muss jedoch auch die Eigenschaften von VRUs berücksichtigen, die bei der Berücksichtigung der V2P-Kommunikation in einem V2X-System viele Fragen aufwerfen. Diese Dissertation befasst sich mit drei Aspekten im Zusammenhang mit der Entwicklung eines V2P-Systems. Der erste Aspekt betrifft die systematische Konzeption eines V2P-Systems nach einem ganzheitlichen Ansatz. Diese Dissertation schlägt einen V2P-Entwurfsrahmen vor, der auf verschiedenen Eingangsgrößen basiert, die für die Entwicklung eines effektiven V2P-Systems erforderlich sind. Dieser Entwurfsrahmen verbessert das Verständnis der V2P-Systemanforderungen und trägt dazu bei, den Entwurfsprozess systematischer zu gestalten. Der zweite Aspekt betrifft die Leistung des V2X-Netzes, wenn eine große Anzahl von VRUs präsent ist. Diese Dissertation schlägt hierfür MC-COCO4V2P vor, einen energieeffizienten Clustering-Mechanismus für Fußgänger zur Eindämmung der Netzüberlastung. MC-COCO4V2P verbessert die Netzleistung, indem die Anzahl der von Fußgängern generierten Sicherheitsmeldungen reduziert wird. Damit wird zudem die Batterielebensdauer der von den Fußgängern genutzten Geräte verbessert. Der dritte Aspekt betrifft die Zuverlässigkeit der Kommunikation zwischen einem Fahrzeug und einem Fußgänger, die kurz vor einem Zusammenstoß stehen. Diese Dissertation stuft eine so wichtige Kommunikation als diejenige ein, die selbst beim Austausch anderer kritischer Sicherheitsnachrichten die höchste Priorität bekommt. Es wird ein Mechanismus vorgeschlagen, der es den umgebenden Verkehrsteilnehmern ermöglicht, ihre Kommunikationspriorität vorübergehend zu verringern. Dies führt zu einem bevorzugten Medienzugriff für die durch eine Kollision gefährdeten Verkehrsteilnehmer, was zu einer höheren Dienstgüte (QoS) für deren Kommunikation führt.

Contents

1 Introduction **1**
1.1 Motivation 2
1.2 Problem Statement 3
1.3 Key Objectives and Contributions 5
1.3.1 Objectives 5
1.3.2 Contributions 5
1.4 Outline 6

2 Background **9**
2.1 Vehicle-to-Everything Communication 10
2.1.1 Vehicle-to-Vehicle Communication 10
2.1.2 Vehicle-to-Infrastructure Communication 10
2.1.3 Vehicle-to-Network Communication 11
2.1.4 Vehicle-to-Pedestrian Communication 11
2.2 V2X Applications 11
2.2.1 Traffic Safety 12
2.2.2 Traffic Efficiency 12
2.2.3 Infotainment 12
2.3 V2X Communication Technologies 13
2.3.1 Dedicated Short Range Communication 13
2.3.1.1 Physical Layer 13
2.3.1.2 Medium Access Control Layer 14
2.3.1.3 Network and Transport Layer 17
2.3.1.4 Application Layer 24
2.3.1.5 Security 24
2.3.1.6 Safety Messages 25
2.3.2 Cooperative ITS 26
2.3.2.1 Access Layer 27
2.3.2.2 Network and Transport Layer 29
2.3.2.3 Facilities Layer 29

2.3.2.4 Application Layer 31
2.3.2.5 Safety Messages 32
2.3.3 Cellular-V2X 33
2.3.3.1 Communication Modes 33
2.3.3.2 V2X Application Server 34
2.3.3.3 Quality of Service Provision 35
2.3.3.4 Authorization 35
2.4 Vehicular Ad-hoc Networks 35
2.4.1 Characteristics of VANETs 35
2.4.2 Clustering in VANETs 36
2.4.2.1 Overview 36
2.4.2.2 Related Work 37
2.5 V2X Pilot Deployments 38
2.5.1 North America 39
2.5.2 Europe 39
2.5.3 Asia 40
2.6 Discussion 40

3 Vehicle-to-Pedestrian Communication Systems 41
3.1 V2P Introduction 41
3.2 Fundamentals of V2P Crash Prevention System 42
3.2.1 Contact Duration 42
3.2.2 Phases of Crash Prevention 44
3.2.3 V2P System Architecture 45
3.3 VRU Characteristics 46
3.4 V2P System Requirements 47
3.4.1 V2P System Design 47
3.4.1.1 Overview 47
3.4.1.2 Related Work 47
3.4.1.3 Analysis 49
3.4.2 Network Congestion Control 50
3.4.2.1 Overview 50
3.4.2.2 Related Work 50
3.4.2.3 Analysis 53
3.4.3 Quality of Service for Crucial V2P Communication 54
3.4.3.1 Overview 54
3.4.3.2 Related Work 54

3.4.3.3 Analysis . . . 56
3.5 Evaluation of V2P Communication . . . 56
3.5.1 Scenario . . . 57
3.5.2 Simulation Environment . . . 58
3.5.3 Evaluation . . . 59
3.5.3.1 Network load . . . 59
3.5.3.2 Quality of Service for Crucial V2P Communication 62
3.6 V2P Research and Development Tools . . . 64
3.6.1 Simulation tools . . . 64
3.6.1.1 OMNeT++, Veins, and SUMO . . . 64
3.6.1.2 Network Simulator-3 . . . 64
3.6.2 Hardware tools . . . 64
3.6.2.1 V2X Hardware . . . 64
3.6.2.2 Proof-of-Concept tools . . . 64
3.7 Discussion . . . 65

4 V2P System Design Framework 67
4.1 Introduction . . . 67
4.2 Design Framework . . . 67
4.2.1 Types of VRUs . . . 67
4.2.2 Pre-crash scenarios . . . 68
4.2.3 Mode of Communication . . . 69
4.2.3.1 Direct mode . . . 70
4.2.3.2 Indirect mode . . . 70
4.2.3.3 Hybrid mode . . . 70
4.2.4 Type of Application . . . 71
4.2.4.1 Safety . . . 71
4.2.4.2 Convenience . . . 71
4.2.5 Communication Technology . . . 71
4.2.5.1 IEEE 802.11p . . . 72
4.2.5.2 Cellular Technology . . . 72
4.2.5.3 Wi-Fi . . . 72
4.2.5.4 Localization using tags . . . 73
4.2.5.5 Bluetooth . . . 73
4.2.5.6 700 MHz ITS Band . . . 73
4.2.5.7 802.15.4 . . . 73

4.2.6 VRU Devices 74
4.2.6.1 Smartphone 74
4.2.6.2 Helmet 74
4.2.6.3 Specialized tag 74
4.2.7 Role of VRU Devices 74
4.2.7.1 Active 75
4.2.7.2 Passive 75
4.2.8 Notification Recipients 75
4.2.9 Response Time Requirement 76
4.3 Evaluation of Design Framework 76
4.3.1 Concept 76
4.3.2 Scenario 78
4.3.3 Simulation Environment 78
4.3.4 Evaluation 79
4.4 Discussion 81

5 Network Congestion Control in V2P Networks 83
5.1 Introduction 83
5.2 Multi-channel Clustering-based Congestion Control 84
5.2.1 Concept 84
5.2.2 Assumptions 85
5.2.3 Clustering Architecture 85
5.2.3.1 Clustering Algorithm 86
5.2.3.2 Cluster Head Selection Criteria 87
5.2.3.3 Cluster-aware Safety Message Delivery 87
5.2.3.4 Cluster Messages 88
5.2.4 Transmission Power Control 88
5.2.4.1 Power consumption without clustering 89
5.2.4.2 Power consumption with clustering 89
5.3 Performance Evaluation 91
5.3.1 Scenario 91
5.3.2 Simulations 93
5.3.3 Evaluation 94
5.4 Discussion 98
5.4.1 Cluster parameters 98
5.4.2 Channel modeling 99
5.4.3 Communication technology for pedestrian collaboration 99

5.4.4 Infrastructure-based V2P approaches 99
5.4.5 Reliability of intra-cluster and inter-cluster communication . 100
5.4.6 Accuracy of the sensor data 100
5.4.7 Security and privacy 100

6 On-demand Quality of Service for Crucial Communication 101
6.1 Introduction 101
6.2 System Model 101
6.2.1 Concept 102
6.2.2 Priority Request Message 102
6.2.3 Processing of Priority Request Message 104
6.2.3.1 Priority Request Message on Transmitter Side . . . 104
6.2.3.2 Priority Request Message on Receiver Side 104
6.3 Performance Evaluation 105
6.3.1 Scenario 105
6.3.2 Simulation 106
6.3.3 Evaluation 107
6.4 Discussion 112

7 Summary 113
7.1 Research Contributions 113
7.2 Practical Challenges 114
7.3 Future Work 115

Bibliography 117

List of Figures 135

List of Tables 137

Abbreviations 139

1 Introduction

Intelligent Transportation Systems (ITS) promise to improve the safety and efficiency of traffic. ITS use Information and Communication Technologies (ICT) in order to establish communication among various entities enabling appropriate decision-making. For example, as part of ITS, the vehicle may broadcast its current location to other vehicles in order to make them aware of itself. Various entities that are part of ITS are cars, freight vehicles, road infrastructure, pedestrians, bicyclists, etc. One of the primary objectives of ITS is to create a dynamic and localized ecosystem of ITS entities through information-sharing in order to improve safety. Such information sharing among vehicles is called as Vehicle-to-Vehicle (V2V) communication. V2V communication systems have been standardized in order to enable seamless information sharing among vehicles from different manufacturers. As part of V2V communication, vehicles may broadcast so-called safety messages periodically containing information about their current status, such as location, speed, and direction. Extending V2V communication further, Vehicle-to-Infrastructure (V2I), Vehicle-to-Network (V2N), and Vehicle-to-Pedestrian (V2P) communication have also been proposed. Collectively, such systems are called as Vehicle-to-Everything (V2X) communication system.

Pedestrians, bicyclists, and Motorized Two-Wheelers (MTWs), also known as Vulnerable Road Users (VRUs), are one of the weakest components of ITS from a safety perspective. However, with the advent of new communication technologies and ubiquitous smartphones, VRU protection may no longer be dependent solely on the vehicle's safety systems. VRUs may also participate in the ITS information-sharing in order to increase awareness of their presence. Such VRU participation establishes communication among vehicles and VRUs which is referred to as V2P communication. Although V2V and V2I systems have been well-researched, the V2P system is a relatively new concept and has gained the attention of researchers only in the last few years. It is possible for V2P systems to leverage present V2V communication standards. However, they still need to consider the differentiating characteristics of VRUs, such as mobility, communication device, VRU density, etc. Therefore, a V2P system model needs to be developed which is adapted to the VRU characteristics.

This dissertation deals with the V2P communication system aspect of the integration of VRUs into ITS. It proposes a framework for V2P communication that enables scalability, reliability, and lowering the impact on present V2V communication.

1.1 Motivation

VRUs are the least protected road entities in the crash and hence their fatality rates are high. Despite the increasing penetration of modern Advanced Driver Assistance Systems (ADAS) in vehicles, there were 10,386 and 1605 VRU fatalities in the USA and Germany in 2012, respectively [Int14]. Also, different countries have varying VRU fatalities rates, possibly due to differing lifestyles and different urban planning. Figure 1.1 shows such VRU fatality rates by types and countries [Int14]. As we can see, the USA has higher pedestrian fatalities and lower cyclist

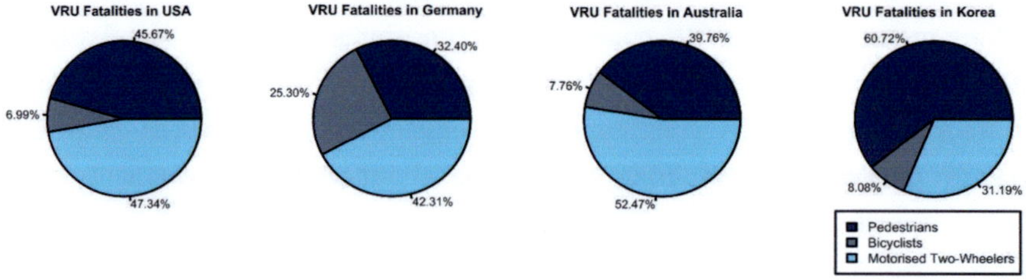

Figure 1.1: VRU fatalities by type and country

fatality rates compared to Germany. Similarly, fatality rates of pedestrians are much higher in South Korea as compared to Australia. These differences have forced the research community, authorities, and technology developers around the world to prioritize the research and development of VRU protection measures. Various pilot projects have been undertaken in the USA and Europe in order to kick-start the development of VRU protection technologies [VRU16, InD18, Aut17, XCY18].

Initially, it was thought that self-driving vehicles may eliminate human errors and hence reduce all types of crashes. Self-driving vehicles deploy a range of sensors, such as cameras, LIDAR, and RADAR, in order to detect crash possibilities. However, they are limited by their requirement of Line-of-Sight (LOS) and inoperability in adverse weather and low light conditions. There have already been a few incidents where self-driving cars have crashed into other vehicles highlighting the need for the deployment of collaborative V2V safety communication. However, a fatal crash in 2018, which involved a self-driving vehicle and a pedestrian-cyclist, has boldly underscored the limitations of these technologies further [Gua18] and has created an urgent need to research and develop a collaborative VRU protection system.

Figure 1.2: Dashcam footage of a self-driving vehicle just before fatal crash [Gua18]

Some efforts have already developed proof-of-concepts of V2P communication based collaborative VRU protection system [WMY+14, DSCP14, HC17, AMS+14, BSN14, TSKF+15]. However, these efforts design only a communication mechanism and lack a rounded approach towards the V2P system design. This dissertation employs a holistic approach with various aspects of the V2P system in mind, which include communication mechanisms, different VRU characteristics, effects of V2P communication on the present V2V system and also, Quality-of-Service (QoS) of communication for a pair of vehicle-VRU that is on the verge of collision.

1.2 Problem Statement

As V2V systems have been extensively researched and standardized, naturally, V2P system will inherit many of their characteristics. However, the V2P system is more than just a communication mechanism among vehicles and VRUs. VRUs differ from vehicles in many aspects and hence effective V2P system design must consider these aspects. The various aspects of V2P design challenges can be summarized below:

1. VRUs vary in their characteristics, such as mobility, pre-crash scenarios, traveling patterns. Also, VRU devices, such as smartphones, may perform different operations while VRUs are using the road and hence, VRUs may not

always be warned about the potential collision reliably. These constraints demand that V2P safety communication systems need a different approach as compared to existing V2V safety communication systems. This requires various VRU characteristics to be researched and pre-crash scenarios to be identified.

2. VRUs can be present in large numbers, especially in urban areas. When V2P communication is enabled, it may affect the existing V2V safety communication. The impact of V2P on V2V, when large number of VRUs are present, needs to be evaluated.

3. A good crash detection system detects the potential crash sufficiently early to allow enough time for braking. This also applies to a V2P communication system. It should allow sufficient reaction time after the crash is predicted. This requires research and evaluation of various VRU pre-crash scenarios under various communication mechanisms. Various crash metrics, such as first contact and available response time, need to be calculated to determine appropriate communication mechanisms.

4. A good V2P system needs to minimize the impact on V2V. The large number of VRU-generated safety messages may congest the network and affect the safety-critical V2V communication adversely. A network congestion control mechanism is needed which would reduce the network load caused by VRU-generated safety messages.

5. When a vehicle and a VRU are on the verge of a collision, they need to be able to communicate reliably in order to avoid the crash. In presence of a large number of VRUs, the VRU-generated safety messages may affect such crucial V2P communication. A mechanism, which would detect the presence of such crucial communication and guarantee its reliability, is needed.

6. VRU devices, such as smartphones, are power constrained. A mechanism is needed which optimizes the power consumption while allowing VRUs to participate in the collaborative safety communication.

1.3 Key Objectives and Contributions

1.3.1 Objectives

This dissertation focuses on the communication protocol design aspects and a Quality-of-Service framework for a V2P communication system. The main objectives are as follows:

1. Investigate various VRU characteristics, pre-crash scenarios, and phases of a crash detection system.
2. Examine the effects of V2P communication on existing V2V network as well as crucial V2P communication under high VRU density.
3. Determine the design input parameters for a V2P system and propose a framework for designing a V2P communication mechanism. Evaluate the framework with various input parameters to provide insights into the effects of input parameters on V2P system performance.
4. Determine whether clustering of VRUs can reduce the network congestion caused by VRU-generated safety messages and whether a dynamic power control mechanism during the clustering process can optimize the power consumption.
5. Determine whether on-demand QoS can be provided for crucial V2P communication by requesting higher priority.

1.3.2 Contributions

This dissertation has made the following contributions towards its objectives:

1. A review and analysis of various road safety statistics, VRU characteristics, pre-crash scenarios, and other relevant approaches [SS19b].
2. Development of a V2P communication design framework which employs a holistic approach towards V2P system development. This framework provides various design considerations in order to provide a structured approach towards system development [SS19b].
3. Provided V2P design framework research results to European Telecommunications Standards Institute Special Task Force 565 (ETSI STF 565) towards the development of VRU service specification [SS19a].

4. Development of a clustering algorithm for network congestion control. The congestion control mechanism is technology-agnostic and can be employed using any technology that supports ad-hoc communication. The clustering algorithm employs dynamic power levels to achieve the energy efficiency [SS20].

5. Development of a special network message that can be broadcast by a crucial pair of vehicle and VRU in order to request higher priority for their communication [SSS20].

All new components have been evaluated using the OMNeT++, Veins, and Simulation of Urban Mobility (SUMO) simulation environment.

1.4 Outline

This document is set out as follows:

Chapter 2 provides a background of V2X communication, V2X standards, and various aspects of V2X communication. It also provides an overview of V2X pilot projects undertaken in multiple countries.

Chapter 3 presents the fundamentals of V2P communication and discusses the state of the art and current challenges of the V2P systems. It also evaluates the V2P network performance underscoring the need for solutions to these challenges.

Chapter 4 presents the V2P design framework with the underlying design choices. It discusses the various design inputs that the design framework is based on and shows their codependency by undertaking a case study for different configurations of the V2P system.

Chapter 5 presents a novel clustering algorithm, called MC-COCO4V2P, which is a network congestion control mechanism for V2P networks. Various elements of the algorithm are discussed and the proof of improvement in network performance is presented.

Chapter 6 presents an on-demand Quality-of-Service mechanism for the crucial communication between a pair of a vehicle and pedestrian. The details of the mechanism are discussed and the improvements in the reliability of the crucial communication are demonstrated.

The document concludes with Chapter 7 which provides the summary of the research contributions, practical challenges, and future work.

2 Background

Automotive safety systems have evolved greatly over the last few decades. Passive safety systems, such as seat-belts, airbags, have been commonly deployed in vehicles to protect passengers after the crash. Alternatively, active safety systems have been deployed to assist the drivers to avoid the crash in the first place. Anti-lock Braking System (ABS), Electronic Stability Control (ESC) are some common examples of active safety systems. Improvements in communication and sensing technologies have paved the way to more advanced active safety systems, also known as Advanced Driver Assistance Systems (ADAS).

ADAS comprises various sensors, such as Light Detection and Ranging (LiDAR), RAdio Detection And Ranging (RADAR), camera, etc. in order to assist drivers in avoiding crashes. However, these technologies are constrained by limited Field-of-View (FOV), Line-of-Sight (LOS) requirement, low number of object-tracking capabilities, and also, lowered effectiveness in low light and adverse weather conditions. A solution, which would enable vehicles to be aware of their surroundings under such conditions is needed. Vehicle-to-Everything (V2X) communication was proposed to be part of this solution. V2X overcomes the limitations of existing ADAS technologies by providing 360° of awareness, Non Line-of-Sight (NLOS) operation capability, increased number of trackable objects, and better performance in adverse weather. It equips vehicles with a communication mechanism that enables them to communicate with their surroundings. The communication involves information-sharing about the vehicle's current status, such as location and speed. This information is then further processed by the receiving surrounding entities which help them make better decisions. This results in improved vehicular safety and traffic efficiency performance.

In 1999, Federal Communication Commission (FCC) in the USA allocated 75 MHz of spectrum (between 5.850 - 5.925 GHz) for vehicular safety communication. This formed the basis of current V2X technologies in the USA and Europe.

2.1 Vehicle-to-Everything Communication

V2X communication involves vehicles sharing their current status to surrounding entities. Their surrounding may involve other vehicles, infrastructure units, network (manufacturer, operator, or cellular), and VRUs. This shared information may then be used for safety and efficiency applications.

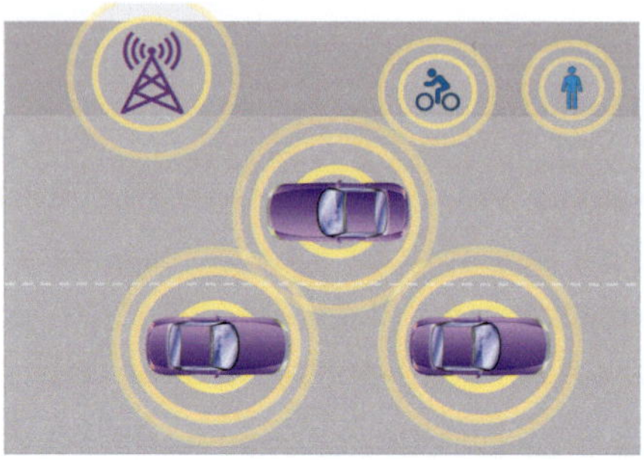

Figure 2.1: Vehicle-to-Everything communication illustration

2.1.1 Vehicle-to-Vehicle Communication

Vehicle-to-Vehicle (V2V) communication involves two or more vehicles sharing their status periodically. The vehicle may broadcast its status up to 10 times per second (10 Hz periodicity). The status is broadcast using a special message called as *safety message* or *beacon*. The process of transmitting safety messages periodically is also called beaconing. This helps enhance their visibility to other vehicles, predict their trajectory, and helps avoid any potential crash. For example, when braking, the vehicle may broadcast its braking status to the following vehicles which then may also decide to apply brakes based on the stopping distance. Vehicles need to be equipped with the so-called On-Board Unit (OBU) in order to communicate with each other.

2.1.2 Vehicle-to-Infrastructure Communication

Vehicle-to-Infrastructure (V2I) communication involves the information sharing between vehicles and infrastructure nodes for safety and efficiency purposes. Infrastructure nodes are the fixed nodes that are deployed roadside and also called as Road-Side Units (RSUs). RSUs may serve various purposes in safety as well as efficiency communication. For example, an RSU may broadcast the current phase of the traffic light to the surrounding vehicles which may help them determine appropriate speed resulting in improvement in traffic efficiency. RSUs may also be equipped with other communication technologies, such as cellular or Wi-Fi, enabling them to gather information from various stakeholders which can then be forwarded to vehicles. Deployment of RSUs may play a key role in the development of smart infrastructure.

2.1.3 Vehicle-to-Network Communication

Vehicle-to-Network (V2N) communication involves an exchange of information between vehicle and backend cloud of a manufacturer or fleet operator over the cellular network for efficiency. The vehicle may send various types of non-critical information, such as, its current location, vehicle diagnostics information, data gathered while traveling, etc., to the backend. This information is then analyzed in order to improve operational efficiency.

2.1.4 Vehicle-to-Pedestrian Communication

Vehicle-to-Pedestrian (V2P) communication involves information sharing among vehicles and VRUs for safety and efficiency applications. As part of safety applications, vehicles and VRUs exchange safety messages among themselves periodically. The vehicle's OBU may communicate with VRU device, such as smartphone, to detect and predict the movements in order to calculate the crash probability. If a crash is predicted, then the vehicle driver and VRU may be warned through haptic, audio, or visual mechanisms. V2P safety applications may have stricter communication requirements in terms of latency and periodicity of exchange of safety messages as compared to efficiency applications.

V2P efficiency applications involve providing various non-critical services to VRUs. Examples of such services are nearby accident information, current traffic status, providing real-time location of public transport vehicles, etc.

2.2 V2X Applications

V2X applications can be classified into three major categories. They are discussed below.

2.2.1 Traffic Safety

V2X safety applications are primarily focused on reducing crashes through the dissemination of critical information in real-time. Vehicles may broadcast the safety messages to their surrounding entities containing their current status. This may include various elements, such as their speed, location, current brake status, path history, and path prediction. These elements are then used by other vehicles and infrastructure nodes for various applications. Examples of such applications are hazard warning, cooperative collision warning, emergency brake warning, etc. These applications are able to predict the crash situations and warn the driver in advance to avoid the crash. In the case of self-driving vehicles, such communication may be used for coordinated driving.

As safety applications rely on critical information being delivered in real-time, they have stringent requirements in terms of latency, packet delivery reliability, safety message periodicity, and communication range. Delivery of information for safety-critical applications takes the highest priority among all categories of V2X communication.

2.2.2 Traffic Efficiency

V2X efficiency applications are focused on improving traffic management. They enable non-critical information sharing in order to improve the traffic flow which results in improved road congestion, reduced travel time, and less fatigue for drivers. Efficient traffic management also improves the fuel economy of vehicles reducing the impact on environment. Some examples of V2X efficiency applications are in-vehicle signage for various road signs, Green Light Optimal Speed Advisory (GLOSA), traffic light phase notification, etc.

As V2X efficiency applications are not safety-critical, their information has lower priority as compared to safety applications. However, this information is still time-sensitive and should be delivered within the defined duration.

2.2.3 Infotainment

Infotainment applications are focused on improving the experience of passengers. These applications may provide targeted information, such as point of interest, local product advertisements, parking information, etc. Information delivery for these applications takes the lowest priority among all V2X application categories.

2.3 V2X Communication Technologies

Dedicated Short Range Communication (DSRC) and Cooperative ITS (C-ITS) technologies have been developed in the USA and Europe respectively to enable V2X communication. DSRC and C-ITS technologies have been extensively researched and also, have been deployed as part of the pilot projects. Cellular-V2X (C-V2X) is another new technology that has been proposed as part of 5G. In this section, we look at the various V2X technologies.

2.3.1 Dedicated Short Range Communication

DSRC is based on Wireless Access in Vehicular Environment (WAVE) protocol architecture. IEEE 1609 group has standardized various parts of the WAVE protocol stack. Figure 2.2 shows the architecture of WAVE system. We discuss various components of the WAVE protocol stack in this section.

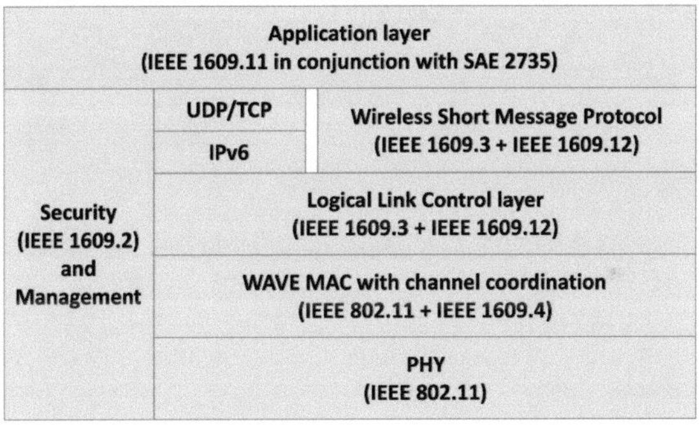

Figure 2.2: DSRC WAVE architecture [IEE19]

2.3.1.1 Physical Layer

WAVE system uses IEEE 802.11p for its physical (PHY) and Medium Access Control (MAC) layer. IEEE 802.11p is a part of the IEEE 802.11 standard and is specially designed to support the high mobility outdoor environment of V2X communication. 802.11p-compliant devices are able to operate outside the context of Basic Service Set (BSS) [IEE16d]. In the case of legacy 802.11 devices, it requires the BSS to be established which causes communication delay. By eliminating this requirement, 802.11p allows ad-hoc communication without any delay even in the case of highly mobile vehicles. This feature is enabled by setting the parameter *dot11OCBActivated* to *true*.

802.11p is supported by the Orthogonal Frequency Division Multiplexing (OFDM) PHY layer system in half-clocked operation mode. It provides 10 MHz channel spacing and supports data rates from 3 to 27 Mb/s. This system uses Binary Phase Shift Keying (BPSK) or Quadrature Phase Shift Keying (QPSK) modulation for lower data rates and 16- or 64-Quadrature Amplitude Modulation (16-QAM or 64-QAM) for higher data rates.

2.3.1.2 Medium Access Control Layer

WAVE system's MAC layer is standardized using IEEE 802.11p and IEEE 1609.4 for MAC layer operations and multi-channel coordination, respectively. Figure 2.3 shows the channel allocation for DSRC technology in USA [IEE19].

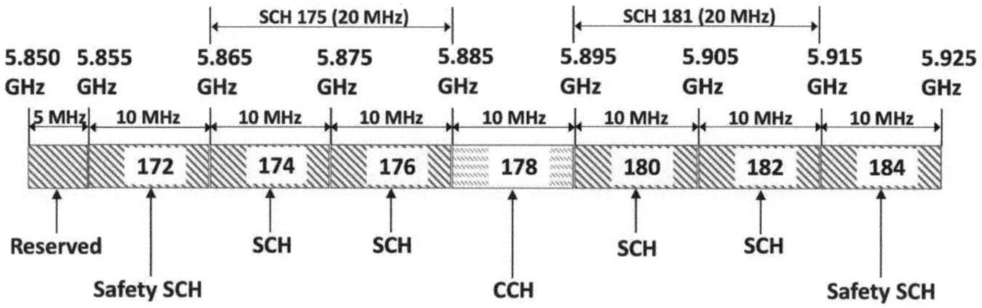

Figure 2.3: 802.11p channel allocation for DSRC

Channel Allocation

The 5 MHz band between 5.850 MHz - 5.855 is reserved for future use. The rest of the 70 MHz band is divided equally into 7 channels of 10 MHz each. Channel 178 is allocated as Control Channel (CCH). CCH may be used for safety and non-safety communication. Channel 172, 174, 176, 180, 182, and 184 are allocated as Service Channels (SCHs). Out of these, SCH 172 and 184 are dedicated for public safety communications involving V2X safety-critical communication. SCHs 174, 176, 180, and 182 can be used for non-critical communication. Also, channels 174 and 176 may be combined to create channel 175 with 20 MHz bandwidth. Similarly, channels 180 and 182 may be combined to create channel 181 with 20 MHz bandwidth. From the higher layers, only Wave Short Message Protocol (WSMP) packets and system management messages are allowed to be transmitted on CCH. SCHs can be used for non-safety messages, which include general-purpose WSMP, IPv6, and management messages.

Channel Interval

IEEE 1609.4 standardizes the channel coordination for WAVE architecture. The channel access is coordinated by providing various mechanisms in the time and frequency domain. Channel access is synchronized using Coordinated Universal Time (UTC) which is provided by Global Positioning System (GPS) [IEE19]. To coordinate the access, a time interval parameter called as *time slot* is defined. A *sync interval* consists of two such consecutive time slots, *time slot 0* and *time slot 1*. WAVE architecture standardizes the parameter value of both time slots to 50 ms making the *sync interval* 100 ms long [IEE16a]. A *guard interval* is a time interval at the beginning of each time slot which is used to accommodate the channel switching and synchronization buffer time. WAVE architecture requires the *guard interval* parameter value to be 2 ms. Figure 2.4 illustrates the channel interval concept.

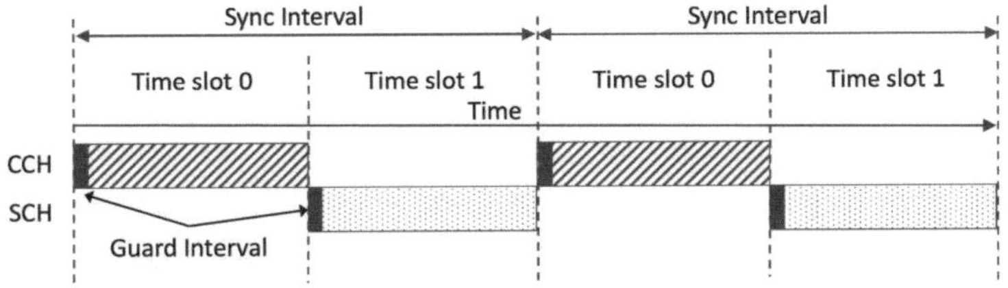

Figure 2.4: Channel synchronization and intervals [IEE16a]

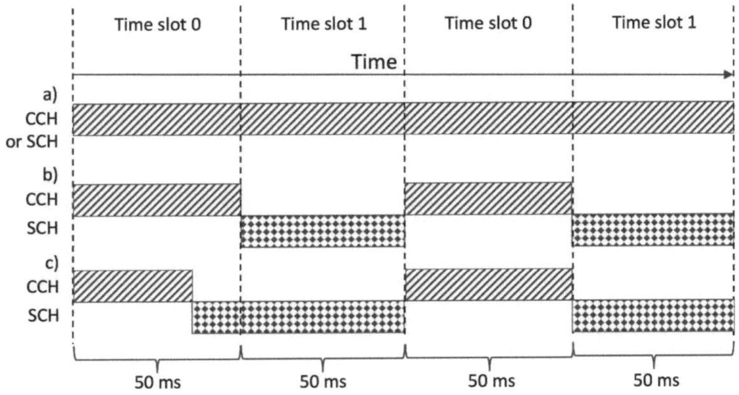

Figure 2.5: Channel access: (A) Continuous, (B) Alternating, and (C) Immediate [IEE16a]

Channel Coordination

WAVE-compliant devices may access different channels during the different channel intervals. Figure 2.5 shows different possibilities of channel coordination options.

During continuous channel access, the device may be tuned to either CCH or SCH continuously. Alternating access involves the device being tuned to different channels during consecutive time slots. For example, the device may be tuned to CCH in *time slot 0* followed by SCH in *time slot 1*. Immediate access mode supports the device to switch to a different channel immediately for longer than one-time slot duration. Multi-radio WAVE devices may choose different channel coordination mechanisms for individual radios.

Quality of Service Provision

WAVE MAC layer supports the Carrier Sense Multiple Access/Collision Avoidance (CSMA/CA) mechanism to control access to the channel. It also supports Enhanced Distributed Channel Access (EDCA) to support different QoS for various types of network packets. EDCA mechanism differentiates the packets into 4 different categories, called Access Categories (ACs), which can be seen in table 2.1. The priority of the packet increases with the *AC Index* (ACI). EDCA

Description	**AC Index**	**AC**
Best Effort	AC_BE	0
Background	AC_BK	1
Video	AC_VI	2
Voice	AC_VO	3

Table 2.1: Access categories in EDCA [IEE16d].

supports a mechanism called a Contention Window (CW) for each AC. This CW[AC] controls the duration of a backoff timer which is different for each AC. Contention window size is controlled by its minimum and maximum bounds which are specified using parameters aCWmin and aCWmax. EDCA mechanism also defines Arbitration Interframe Space Number (AIFSN) which is the amount of time a device probes the channel for the idle state before the start of transmission. Table 2.2 shows the 802.11p-specific CW and AIFSN parameters as defined in the standard IEEE 802.11 [IEE16d]. Higher layers set the User Priority (UP) field on a per-message basis which is then used for deciding EDCA priority of the packet at MAC layer [IEE19].

AC	AIFSN	CWmin	CWmax
AC_BE	9	aCWmin	aCWmax
AC_BK	6	aCWmin	aCWmax
AC_VI	3	(aCWmin+1)/2–1	aCWmin
AC_VO	2	(aCWmin+1)/4–1	(aCWmin+1)/2–1

Table 2.2: EDCA QoS parameters [IEE16d].

Channel Routing

Similar to the UP, higher layers in WAVE architecture can also set channel, data rate, and transmission power level on a per-message basis. This requires a channel routing mechanism to inspect every request coming from a higher layer, map the UP to EDCA priority, and forward it to the appropriate EDCA queue. Figure 2.6 illustrates the functioning of different WAVE MAC components in tandem.

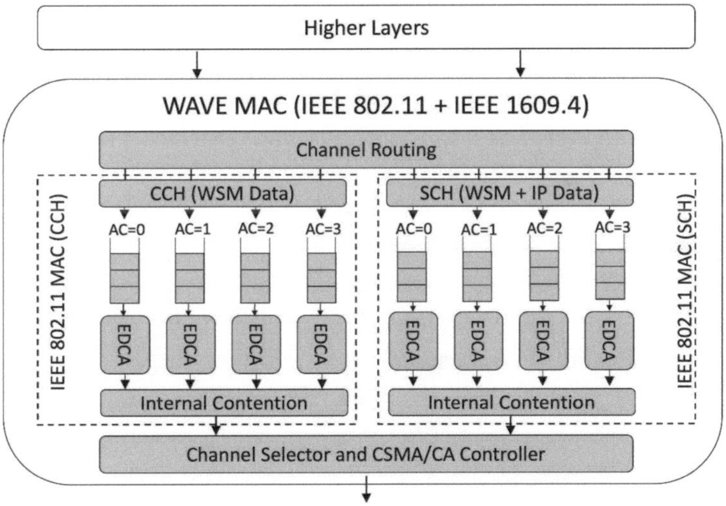

Figure 2.6: WAVE MAC illustration

2.3.1.3 Network and Transport Layer

IEEE 1609.3 standard specifies the network and transport layer of the WAVE architecture. These layers are jointly called WAVE Networking services. Figure 2.7 illustrates the WAVE networking services components. WAVE Networking services are divided further into two functionalities which are discussed below.

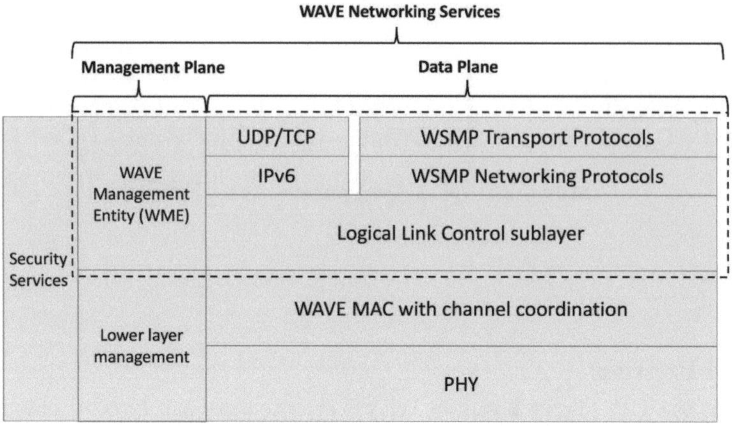

Figure 2.7: WAVE Networking Services

A. Data Plane

WAVE Networking services allow higher layers to set various transmission parameters, such as data rate, MAC channel, transmit power, and time slot per-message basis while WSMP is being used. For IP-based messages, these transmission parameters are stored as a transmitter profile. The data plane consists of the following components:

1. **Logical Link Control sublayer**
   Logical Link Control sublayer (LLC) provides services to the WSMP and IPv6 layer. It uses *EtherType* parameter to differentiate between the WSMP and IPv6-based messages. Higher layers (WSMP or IPv6) request the LLC services using DL-UNITDATA.request (IPv6 layer) or DL-UNITDATAX.request (WSMP layer). Upon receiving the packet from WSMP or IP layer, LLC sets the *Type* field to appropriate *EtherType* value and passes the packet to lower layer using MA-UNITDATA.request (for IPv6) or MA-UNITDATAX.request (for WSMP) service primitives. When LLC receives the packet from the lower layer, it reads the *EtherType* value and then forwards it appropriately (to WSMP or IPv6 layer).

2. **WAVE Short Message Protocol**
   WAVE Short Message Protocol (WSMP) exchanges WAVE Short Messages-(WSMs) for communication. WSMP provides services to the application layer as well as WAVE Management Entity (WME) layer and both

are considered as higher layers. Higher layers send a service primitive WSM-WaveShortMessage.request to invoke WSMP services. WSM-WaveShortMessage.request primitive may contain the following elements:

- Peer MAC address
- Data
- User Priority
- Channel Identifier
- Time Slot
- Data Rate
- Transmission Power Level
- Channel Load
- Expiry Time
- Provider Service Identifier

Upon reception of valid WSM-WaveShortMessage.request, WSMP prepares a WSM containing the above elements and passes it to LLC using DL-UNITDATAX.request primitive. WSMP header contains *Address Info* field which specifies *Provider Service Identifier (PSID)* in order to identify higher layer entity.

LLC notifies WSMP of a new WSM reception via DL-UNITDATA.indication. Upon reception of a valid WSM, WSMP determines the destination higher layer using *PSID*. It then forwards *Data* to this higher layer from the received WSM.

3. **Internet Protocol version 6**
   WAVE architecture supports Internet Protocol version 6 (IPv6) as one of the modes of non-safety communication. IPv6 layer may provide services to higher transport layers. For further details of IPv6, IETF RFC 2460 may be referred.

4. **IP-based Transport Protocols**
   WAVE Networking services support TCP and UDP transport protocols over IPv6 for non-safety communication. For further details of TCP and UDP protocols, IETF RFC 793 and IETF RFC 768 may be referred, respectively.

B. Management Plane

WAVE Networking services provide management functions using WAVE Management Entity (WME). WME management functions are listed below:

1. Channel Access Assignment
   Higher layers in WAVE architecture may exchange data on different channels. As shown in Figure 2.5, WAVE device may access channel in multiple ways. WME provides such channel access based on the higher layer requirements. WME is aware of the device capabilities (single-radio, dual-radio, etc.) and makes appropriate channel assignments decisions based on this knowledge.

2. Service requests fulfillment
   WME provides services to higher layers by accepting various service requests. Service requests are meant to facilitate the WAVE device's internal operation. WME supports the following types of service requests:

   - Provider service request - Higher layers use this request to indicate WME that WME may generate WAVE Service Advertisement (WSA) on its behalf and also, provide SCH access. The WAVE device that transmits WSAs assumes the role of a *Provider*. WME may generate different WSAs for different higher layers.
   - User service request - Higher layers use this request to indicate WME that they are interested in a specific application-service. When the WME identifies such a service, it takes appropriate action as mentioned in the service, such as assign corresponding channel access.
   - WSM service request - Higher layer uses this request to indicate the WME that it is interested in receiving WSMs addressed for a specific *PSID*. When a WSM with the matching *PSID* is received, it is then delivered to this higher layer.
   - Channel service request - Higher layers use this request to indicate WME that it requires channel access to a certain channel in a certain time slot (time slot 0 or 1).
   - Timing Advertisement service request - This request is used to indicate WME that Timing Advertisement (TA) be generated and transmitted. This request may also include a request for channel access. TAs are used for synchronization among WAVE devices.

3. WAVE Service Advertisement operation - *WAVE Service Advertisement* (WSA) is a mechanism to announce various services as well as to announce non-service-related information, such as location, channel parameters. WSA can be broadcast with varying periodicity withing the range of 0.2 Hz - 50 Hz. Higher layers may generate *Provider service request* for WME to initiate WSA transmission. Upon reception of *Provider service request* from higher layers, WME generates WSM-WaveShortMessage.request with the following parameters:
   - Peer MAC address
   - Channel Identifier
   - Time slot
   - Data rate
   - Transmit power level
   - Channel load
   - User priority
   - Expiry time
   - WSM data
   - Provider Service Identifier

   These parameters are then used by *Data Plane* to construct WSA for transmission.

   When a WAVE device receives WSA, WME is notified. WME verifies this WSA for its validity. Information from this WSA may then be stored as part of available services which may later be used by higher layer entities.

C. Packet Formats

All the packet formats of WAVE network and transport layer are required to comply with 802.11 standard when being transmitted over the air. As discussed earlier, various layers may use different service primitives in order to communicate the requirements of channel access, time slot etc internally. However, here we discuss only the packet formats being transmitted over the air.

1. WSM
   WSM packet consists of *WSM Header* and *WSM Data*. WSM header is further divided into WSMP-N-Header and WSMP-T-Header. WSM packet format is shown in Figure 2.8 [IEE16b].

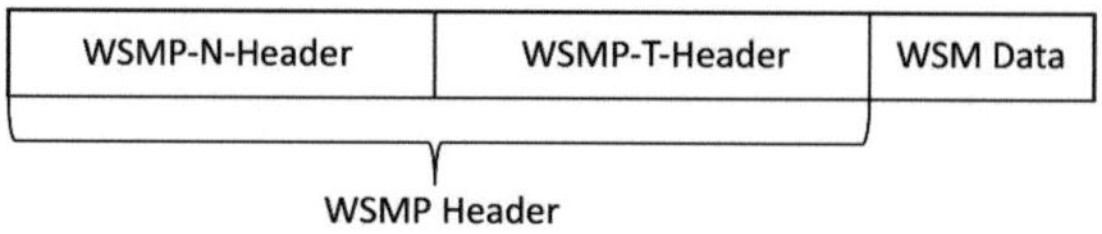

Figure 2.8: WSM packet format

Figure 2.9 shows the WSMP-N-Header format. WSMP-N-Header's *Subtype* field decides the networking protocol features, such as, broadcast, forwarding etc. WSMP-N-Header contains *Transport Protocol ID* (TPID), a field that specifies the WSM-T-Header content. Figure 2.10 shows the WSMP-T-

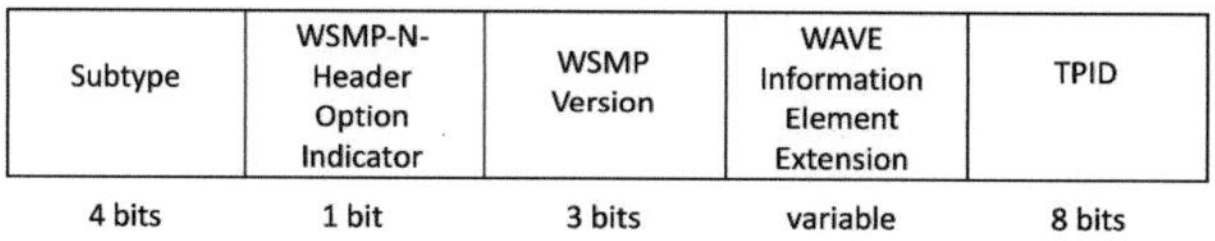

Figure 2.9: WSMP-N-Header format

Header format. Based on the TPID field of the WSMP-N-Header, *Address Info* field of the WSMP-T-Header may take up different formats.

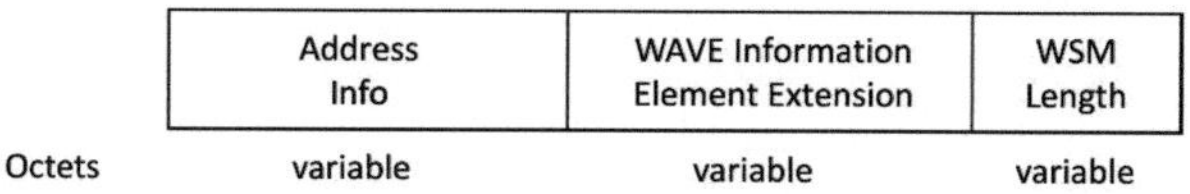

Figure 2.10: WSMP-T-Header format

2. WSA

WSA is encapsulated in the WSM packet as part of WSM data. WSA may be secure (signed) or insecure (unsigned). Figure 2.11 shows the packet format of WSA. WSA packet consists of WSA header and multiple optional fields, such as, *Channel Info Segment, Service Info Segment* and *WAVE Routing Advertisement.* In this section, we describe some of the important fields that are relevant for this thesis. Figure 2.12 shows mandatory fields of WSA header.

- WSA Version - Value of this field is set to 3 and all greater values are reserved for future use [IEE16b].
- WSA Header Option Indicator - This field indicates the presence of optional fields in WSA packet. Format of this field can be seen in the

WSA Header							
WSA Version	WSA Header Option Indicator	WSA Identifier	Content Count	WAVE Information Element Extension	Service Info Segment	Channel Info Segment	WAVE Routing Advert.
4 bits	4 bits	4 bits	4 bits	variable	variable	variable	variable

Figure 2.11: WAVE Service Advertisement format

WSA Version	WSA Header Option Indicator	WSA Identifier	Content Count
4 bits	4 bits	4 bits	4 bits

Figure 2.12: WSA header - mandatory fields

Figure 2.13. Bit value of '1' indicates the presence of the corresponding field.

WAVE Information Element Extension	Service Info Segment	Channel Info Segment	WAVE Routing Advertisement
1 bit	1 bit	1 bit	1 bit

Figure 2.13: WSA header option indicator

- WSA Identifier - This field is used for identification of unique WSA.
- Content Count - WSA recipient uses this field to determine whether the WSA is a repeat of any other previously received WSAs having the same WSA identifier.

3. WAVE Information Element Extension
WAVE Information Element Extension field allows to communicate extra information pertaining to various fields in WSMP and WSA header, and WSA. It consists of various *WAVE Information Elements.* Figure 2.14 shows the packet format of *WAVE Information Element Extension.* Figure 2.15

Count	WAVE Information Elements
variable length	variable length

Figure 2.14: WAVE Information Element Extension

shows the packet format of *WAVE Information Element.* For the unknown *WAVE Element ID*, WAVE device can still process the message and disregard the *WAVE Information Element.* Currently, first 0-23 *WAVE Element ID* values have been allocated for various purpose and 24-255 have been unused.

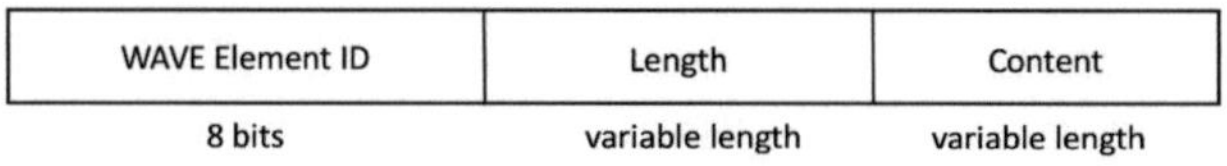

Figure 2.15: WAVE Information Element

2.3.1.4 Application Layer

WAVE architecture supports various safety and non-safety applications that are required for vehicle operation. The applications may pertain to V2I applications, such as, toll collection, or V2V applications, such as emergency warning. However, standardization of such applications is still currently under discussion or undertaken by other organizations. Only toll collection application has been standardized through IEEE 1609.11 as part of the 1609 working group.

The network and transport layer of WAVE architecture identifies various application layers using unique PSIDs. Table 2.3 shows selected PSID value allocation details. Also, the Society of Automotive Engineers (SAE) has standardized the messages, in the standard SAE J2735, that can be exchanged for V2X communication. WAVE Application layers of WAVE devices exchange these messages among each other in order to achieve the goals of the application.

PSID value (hex)	Application	Documentation
0x1	Electronic fee collection	ISO 14906
0x06	Parking management	ISO 15628
0x20	V2V safety and awareness	SAE J2735
0x27	Vulnerable Road Users safety application	SAE DSRC TC
0x82	Intersection safety and awareness	SAE J2735

Table 2.3: PSID Allocations [IEE16c].

2.3.1.5 Security

WAVE architecture's security functionality has been standardized using IEEE 1609.2 [IEE16e]. WAVE architecture supports two types of security services:

1. WAVE Internal Security Services - This service supports security functionality from the physical layer to transport layer. It converts the unsecured protocol data units to secured protocol data units (PDU). The secured PDU may be signed before transmission. Upon reception of the secured protocol data unit, it verifies the authenticity and converts it to the unsecured PDU.

2. WAVE Higher Layer Security Services - This component acts as an interface between the WAVE device and external entities. The external entities may include other WAVE devices or certificate issuing entity. This component is further divided into two components:

   - Certificate Revocation List Verification Entity (CRLVE) - This component verifies the incoming Certificate Revocation List (CRL) and passes it to internal security services.
   - Peer-to-Peer Certificate Distribution Entity (P2PCDE) - This component communicates with peer WAVE devices for exchange of certificates.

WAVE security architecture may use symmetric keys, public-private keys, or digital certificates for security operations.

2.3.1.6 Safety Messages

V2X communication involves exchanging various kinds of information among vehicles for traffic safety and efficiency. The vehicles may exchange their current location, direction, speed, etc. The infrastructure nodes may broadcast the current traffic light phase, respond to green traffic light requests, etc. In order to harmonize such exchange of information, SAE has standardized the message formats in the SAE J2735 document for various types of information [Per17]. WAVE architecture utilizes the SAE J2735 message formats to harmonize the exchange of information. SAE J2735 defines total of 17 message formats for various types of information [Per17]. We discuss selected formats here.

- BasicSafetyMessage (BSM): This is the core V2V safety message. It carries various elements of vehicle's current status. It is divided into two parts:

  1. Part I - This part of the message contains core elements of the vehicle's data viz. location, speed, direction, vehicle size, acceleration, brake status. This is a mandatory part of the message and all elements must be part of the message during its transmission.

2. Part II - This part of the message contains variable optional data elements and the latest events for the elements from Part I, such as activation of Traction Control system (braking system event) [Cro12]. The optional elements are headlight status, wiper status, rain sensor, etc.

BSMs are commonly used in various V2V applications, such as Collision Warning, Emergency Braking, etc.

- PersonalSafetyMessage (PSM): VRUs may broadcast this message to surrounding vehicles to announce their current status. This message format is still being researched.

- Signal Phase and Timing (SPaT) - Infrastructure nodes (RSUs) may broadcast this message to provide information about the current phase of the traffic light. SpaT messages are commonly used to improve traffic efficiency around the intersection.

- Map - This message provides the intersection layout information and is broadcast by RSUs along with the SPaT message.

- SignalRequestMessage (SRM) and SignalStatusMessage (SSM) - SRM is broadcast by vehicles to request priority at the signalized intersections. Common applications include priority requests by public transport vehicles and emergency vehicles. SSM is broadcast by RSUs as a response to SRMs. SSM contains information about the pending priority requests.

2.3.2 Cooperative ITS

Cooperative ITS (C-ITS) system has been developed for V2X communication in Europe. C-ITS has been designed to transform vehicular and infrastructure nodes into ITS stations for traffic safety, efficiency, and infotainment applications. It supports an architecture with multiple modular physical layers (WiFi, Cellular, Bluetooth etc.) that are networked together to support various applications. C-ITS protocol stack has been standardized by European Telecommunication Standards Institute (ETSI). Figure 2.16 shows the reference architecture of C-ITS station [ETS10]. In this section, we discuss C-ITS station, also called ITS-S, components.

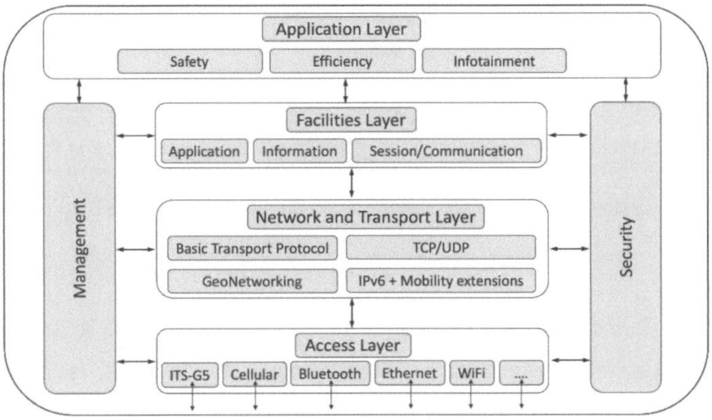

Figure 2.16: C-ITS station reference architecture [ETS10]

2.3.2.1 Access Layer

C-ITS reference architecture conceptually supports multiple physical layers for V2X communication. However, the efforts in Europe have been mainly focused on the development of so-called ITS-G5 technology. ITS-G5 technology is compliant with IEEE 802.11p (part of IEEE 802.11) for its PHY layer [ETS13a]. It relies on IEEE 802.11 and 802.2 for its MAC and data link layer. Figure 2.17 illustrates the ITS-G5 access layer architecture [ETS19]. Similar to DSRC, ITS-G5 too works outside the context of BSS by setting *dot11OCBActivated* to *true* (only for selected bands). ITS-G5 access layer has been divided into 4 frequency bands viz. ITS-G5A, ITS-G5B, IS-G5C, and ITS-G5D. Table 2.4 shows the frequency allocation for the ITS-G5 access layer.

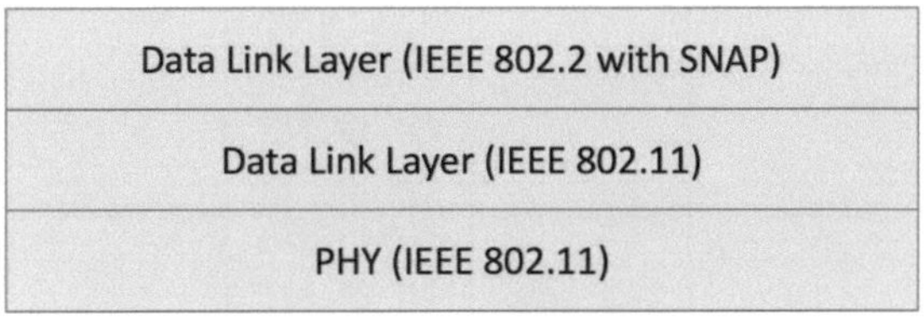

Figure 2.17: ITS-G5 access layer [ETS19]

Channel Allocation

ITS-G5 spectrum is divided into one control channel, called as G5-CCH, and 7 service channels which are called as G5-SCHs. G5-CCH, G5-SCH1, and G5-SCH2 are part of the ITS-G5A band and are dedicated to safety applications. G5-

Band	Frequency Range (MHz)	Usage
ITS-G5A	5875 to 5905	Traffic safety
ITS-G5B	5855 to 5875	non-safety/efficiency
ITS-G5C	5470 to 5725	WLAN, BRAN
ITS-G5D	5905 to 5925	Reserved for future

Table 2.4: ITS-G5 Frequency Allocations [ETS13a].

SCH3 and G5-SCH4 are part of ITS-G5B and are dedicated to traffic efficiency applications. Table 2.5 shows the complete channel grouping for ITS-G5 access layer. Figure 2.18 depicts the channel allocation for ITS-G5. ITS-G5A, ITS-G5B, and ITS-G5D bands operation is supported by setting *dot11OCBActivated* to *true*. However, ITS-G5C band cannot operate when *dot11OCBActivated* is set to *true*.

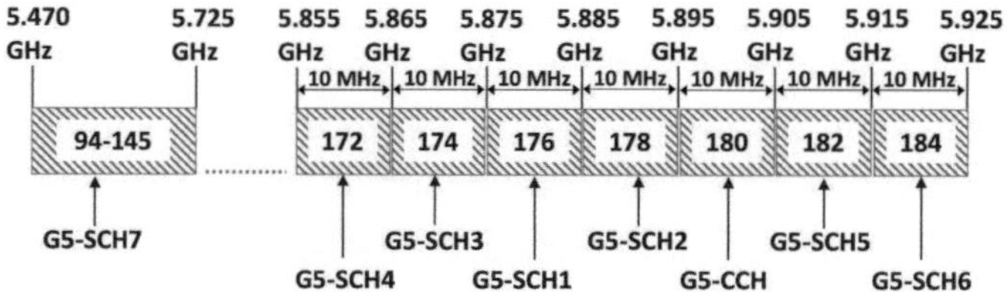

Figure 2.18: ITS-G5 channel allocation

Band	Channels
ITS-G5A	G5-CCH G5-SCH1 G5-SCH2
ITS-G5B	G5-SCH3 G5-SCH4
ITS-G5C	G5-SCH7
ITS-G5D	G5-SCH5 G5-SCH6

Table 2.5: ITS-G5 channel grouping [ETS13a].

Quality of Service Provision

Similar to DSRC access layer, ITS-G5 leverages EDCA functionality of IEEE 802.11 in order to provide QoS.

2.3.2.2 Network and Transport Layer

In this section, we briefly discuss the network and transport layers of C-ITS architecture.

Network Layer

C-ITS architecture supports GeoNetworking and IPv6 protocols at the network layer [ETS14a]. GeoNetworking provides support for ad-hoc networking in desired geographical areas. It supports geographical addressing as well as geographical routing using GeoBroadcast (all nodes in a given area) or GeoAnycast (any node in a given area). GeoNetworking may also support the transport of IPv6 packets. It achieves transport of IPv6 packets using *tunneling* i.e. GeoNetworking packet headers encapsulating IPv6 packets. GeoNetworking protocol stack also provides services to higher layers i.e. transport layer protocols.

Transport Layer

C-ITS architecture supports Basic Transport Protocol (BTP) and TCP/UDP protocols at the transport layer. BTP is a connection-less, end-to-end transport layer protocol [ETS17]. BTP offers non-guaranteed delivery of the packets. In C-ITS architecture, it provides services to the facilities layer and also allows them direct access to lower layer services (GeoNetworking protocol). BTP employs a port mechanism to differentiate between the different facilities layer entities. Table 2.6 lists some commonly known BTP port numbers.

BTP Port number	Facilities layer entity (or application)
2001	Cooperative Awareness
2002	Decentralized Environmental Notification
2005	Service Announcement
2006	Infrastructure to Vehicle Information
2010	Electric Vehicle Charging Spot Notification

Table 2.6: Selected BTP port numbers [ETS16].

2.3.2.3 Facilities Layer

Facilities layer is a middleware layer that provides services to the higher layer entity (ITS applications) [ETS13b]. Facilities layer consists of multiple facilities that are responsible for providing services, functions, or information to the multiple ITS applications. Facilities can be broadly classified into two categories:

1. Common facilities: These facilities are the core service providers that are necessary for operational reliability of the ITS-S and interoperability of the ITS applications. For example, facilities that provide positioning service or time service fall into this category.

2. Domain facilities: These facilities are dedicated to providing services to specific ITS applications. One or more ITS applications may use the same domain facility. However, not all domain facilities are necessary for the operation of an ITS application.

Facilities layer may provide different kinds of support to the application layer. A facility may be classified as one of the following types based on the type of support it provides [ETS13b]:

- Application support: This category includes facilities that provide necessary services for the operation of an ITS application.

- Information support: This category includes facilities that are providers of the common data, such as, maps and database management functionalities, such as, Local Dynamic Map (LDM) for storing the vehicle sensor data and maps.

- Communication support: These facilities provide communication-related functionalities, such as addressing, geocasting, etc.

- Management support: These facilities interface with the management and security entities of the ITS-S.

Figure 2.19 illustrates the facilities layer functionality [ETS09]. Facilities layer also acts as an ITS-S gateway for external systems. These external systems include:

- In-vehicle network

- Road-side infrastructure nodes

- Traffic Management Center

- Back-end networks, such as, Internet

- Personal ITS-S inside the vehicle

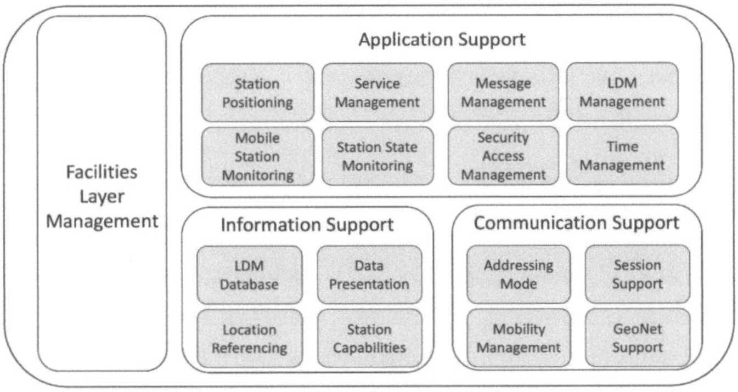

Figure 2.19: C-ITS Facilities layer functionality
[ETS09]

2.3.2.4 Application Layer

C-ITS application layer defines a Basic Set of Applications (BSAs) that help improve traffic safety, traffic efficiency, and other vehicular infotainment services [ETS09]. We list a few selected BSAs below that are identified as the most widely applicable. The end-use cases are grouped together by application which, in turn, are grouped together by *application class.*

1. Cooperative traffic safety:
   - Cooperative awareness - Emergency vehicle warning, intersection collision warning
   - Road hazard warning - Emergency vehicle brake lights, accident notification
2. Cooperative traffic efficiency
   - Speed management - Speed limit notification, traffic light optimal speed advisory
   - Cooperative navigation - Enhanced route guidance and navigation, in-vehicle signage
3. Cooperative local services
   - Location based services - Parking management, Point of Interest notification
4. Global internet services

- Communities service - Fleet management, loading zone management
- ITS station life cycle management - Vehicle software update, vehicle data calibration

2.3.2.5 Safety Messages

Similar to the WAVE protocol stack, C-ITS architecture also defines a set of fundamental building blocks that enable the seamless exchange of different kinds of information among vehicles and infrastructure. However, C-ITS achieves a seamless exchange by defining components at the facilities layer. C-ITS defines Cooperative Awareness Message (CAM) to exchange the information regarding the vehicle's current status, such as location and speed [ETS11]. It also defines a Decentralized Environmental Notification Message (DENM) to exchange the information regarding the vehicle's surroundings, such as traffic conditions [ETS14b]. In this section, we discuss the CAM and DENM mechanisms.

Cooperative Awareness Message

Vehicles equipped with C-ITS technology exchange CAMs among each other to share their current status. CAM contains fields, such as vehicle's speed, position, heading, vehicle length, and vehicle width [ETS11]. CAM operation is managed by the *Message Management* component in facilities layer as shown in Figure 2.19. CAM management entity passes the CAMs received from other vehicles to LDM management. LDM management uses the data in CAMs to update the LDM database containing the information about the surrounding vehicles. CAMs may be generated with different periodicity based on the various use cases. For example, *Emergency Vehicle Warning* use case may require to generate CAMs with 10 Hz periodicity while *Slow Vehicle Indication* use case may generate CAMs with 2 Hz.

Decentralized Environmental Notification Message

ITS-S transmits DENM when it detects a potentially hazardous event (or condition), such as stationary vehicle or collision risk [ETS14b]. DENM contains information about the event, such as type of the event, event location, event detection timestamp, relevant traffic direction, etc. DENM operation is managed by Decentralized Environmental Notification (DEN) basic service residing at *Message Management* in facilities layer. Upon detection of potentially hazardous conditions, DENM is transmitted by ITS-S with certain frequency. This transmission is continued till the ITS-S detects that the event is terminated. To enable this functionality, the following types of DENM have been defined [ETS14b]:

- New DENM: When an ITS-S station detects the hazardous condition for the first time, it generates a new DENM with a new identifier, called as *actionID*.
- Update DENM: As ITS-S transmits DENMs about any event with a certain frequency, it uses this type of DENM to send follow-up DENMs after the first *New DENM* is transmitted.
- Cancellation DENM: This type of DENM is transmitted by the originating ITS-S to notify the termination of the event.
- Negation DENM: This type of DENM is generated by an ITS-S if it detects the termination of an event for which it did not generate the *New DENM* (*New DENM* for the terminated event was generated by other ITS-S).

As DENM information may not be applicable to all the surrounding vehicles, ITS-S may transmit the DENM only to the selected geographical area using the Network layer's GeoNetworking functionality.

2.3.3 Cellular-V2X

Cellular-V2X (C-V2X) is an upcoming technology for V2X communication as part of 5th Generation (5G) cellular networks. Cellular networks till 4G have provided voice and data connectivity to vehicles for infotainment applications that are not safety-critical. However, C-V2X has been designed to support V2X traffic safety applications that have more stringent QoS requirements. In this section, we discuss the basics of C-V2X architecture as being proposed by 3rd Generation Partnership Project (3GPP) standardization body.

2.3.3.1 Communication Modes

C-V2X technology supports V2X communication in two modes:

1. Infrastructure-less mode - In this mode, V2X communication takes place without any infrastructure (base-stations) over the so-called PC5 sidelink interface [3GP19a]. In this mode, the C-V2X device accesses the channel in autonomous mode, also called as *autonomous resource selection* [3GP19b]. This mode uses 64-QAM modulation for communication. PC5 interface provides a connection-less communication i.e. without any signaling overhead. It also supports the exchange of IPv6-based as well as non-IP based V2X messages.

2. Infrastructure mode - In this mode, V2X communication takes place through the infrastructure nodes over the so-called LTE-Uu interface. In this mode, the network schedules the resource access for the C-V2X. This mode supports unicast as well as multicast V2X communication.

When the C-V2X device is within the coverage area, it may use both interfaces. Inside the coverage area, it may operate PC5 interface in *operator managed* mode. Outside the coverage area or in a limited-service area, it may use only PC5 interface with *non-operator managed* mode.

V2X Control Function

V2X Control Function provides functionality related to V2X network operations and resides in the network [3GP19a]. It performs the following functions:

- Provide C-V2X devices with the necessary parameters for V2X communication
- Provide C-V2X devices with the network-specific parameters for V2X communication
- Provide V2X communication parameters for the device operation outside of the network coverage area.

2.3.3.2 V2X Application Server

The V2X application server is responsible for handling the V2X messages from various C-V2X devices. It performs various functions at the back-end [3GP19a]:

- Deliver data to C-V2X devices in a targeted area
- Receive data from C-V2X devices
- Map geographic location information to appropriate cellular infrastructure nodes
- Provide parameters for V2X communication over PC5 interface to V2X Control Function
- Provide parameters for V2X communication over PC5 interface to C-V2X device

V2X Application Server may connect to V2X Control Functions from multiple carrier networks. V2X Control Function of a given carrier network may provide the configuration information for the V2X Application Server to its C-V2X devices.

2.3.3.3 Quality of Service Provision

PC5 interface supports different priority and reliability levels per-packet basis. It uses the Proximity services Per-Packet Priority (PPPP) mechanism to support different priority levels [3GP19a]. Also, it uses Proximity services Per-Packet Reliability (PPPR) mechanism to support different reliability levels. The application layer of the V2X application on a C-V2X device can set the desired priority and reliability levels on a per-packet basis.

For V2X over LTE-Uu link, V2X applications can use certain QoS Class Identifier (QCI) parameter values in order to request the QoS [3GP19a].

2.3.3.4 Authorization

C-V2X device's home carrier network or visitor carrier network is responsible for authorizing the usage of both interfaces [3GP19a].

2.4 Vehicular Ad-hoc Networks

As vehicles communicate with each other using V2X technologies in an ad-hoc manner, their network is called Vehicular Ad-hoc NETwork (VANET). Although the research community initially considered VANETs a subset of Mobile Ad-hoc Networks (MANETs), distinctive characteristics of VANETs have made them dissimilar to MANETs.

2.4.1 Characteristics of VANETs

We discuss the distinctive characteristics of VANETs below:

- Dynamic topology - As vehicles travel at high speeds, the topology of the VANET continuously changes. The topology also depends on the geographical location and the contact duration of the vehicular link is short [KFHG19].

- Dynamic communication environment - As vehicles travel, their surroundings continuously change. In an urban environment, vehicles may be surrounded by structures of various sizes and shapes. Also, vehicle density may be higher in the urban environment. In a non-urban environment, the structure may be absent and the density of vehicles may be lower. These factors affect the vehicular networks [KFHG19].

- Mobility - Mobility of the vehicles is predictable as vehicles follow the roadmap [PIB11].

- Energy Constraints - Vehicles are equipped with enough energy and computing power. These factors are not considered to be constraints [PIB11].

2.4.2 Clustering in VANETs

The clustering process involves organization of a collection of patterns into clusters based on similarity [JMF99]. As vehicle mobility may display recognizable patterns, vehicles can be organized into clusters. These clusters then may be used to achieve various objectives in VANETs. Applications of clustering in VANETs include routing, channel access management, topology discovery, collision avoidance, etc. [CFR+17]. Apart from the clustering efforts for specific applications, there are also significant efforts to design general-purpose clustering algorithms for VANETs [CFR+17].

2.4.2.1 Overview

Any given VANET cluster usually consists of a vehicle node serving as a Cluster Head (CH) and one or more vehicle nodes acting as Cluster Members (CMs). CH and CMs may perform various roles in clustering operations which depends on the application of cluster formation. Clustering in VANETs involves various operational aspects. We discuss these aspects below [CFR+17].

- Neighbor Discovery - In this operation, vehicle nodes broadcast their presence as well as listen to the cluster announcements from nearby vehicles.

- Cluster Head Selection - After the neighbor discovery phase, the vehicle has gathered enough information about its neighboring vehicles. It then may choose an appropriate CH based on the clustering criteria or choose to be CH by itself. Examples of CH selection criteria are the closest neighbor, communication capabilities, distance to the intersection, etc. This criterion is chosen by the clustering system developers.

- Cluster Affiliation - After the vehicle node shortlists appropriate CH, it may send a request to this CH to join the cluster. Once the affiliation request is confirmed by the CH, the node then assumes the CM role.

- Cluster Maintenance - As a CH, the vehicle node is responsible to perform the cluster maintenance operations, such as, poll the cluster members regularly, send cluster awareness announcements, add new cluster members, and remove non-responsive cluster members. As a CM, the vehicle node is responsible to maintain communication with the CH by either responding to the CH's polling messages or sending regular updates to CH to maintain its CM status.

Clustering operation may be carried out using specially designed control messages for cluster-related communication. Also, VANET clustering process may be carried out in either *proactive* or *passive* manner [TMGS13]. If the control messages are exchanged during cluster operation explicitly, then the process is called as *proactive* clustering. In *passive* clustering process, control messages are piggybacked on the data messages.

2.4.2.2 Related Work

VANET clustering is an extensively researched area due to its interesting applications [TMGS13, CFR+17, KA07, AS13, ZSC06, YTH14, ZGZ+19]. In this section, we look at the clustering related work with focus on VANET clustering.

Authors of [TMGS13] propose an cluster-based approach for intersection collision avoidance. This approach is a *proactive* approach that uses Wi-Fi to exchange cluster control messages among vehicles while LTE is used by the Cluster Head to transmit the safety messages on behalf of the cluster. Yang et al. [YTH14] propose a clustering protocol for VANETs to improve the reliability of communication and to provide application-sensitive QoS. This approach uses WAVE architecture's multi-channel operation capability to provide TDMA-based channel access for cluster members. Allouche et al. [AS13] propose a clustering algorithm for beacon-dissemination in a highway scenario. In this approach, CHs are responsible for aggregating messages from CMs, communicating with other CHs, and disseminating information within the clusters. This information is then used to create accurate maps of the vehicle's surroundings. Authors of [MBM11] propose a general-purpose clustering algorithm based on location and direction. This approach forms clusters near the intersection area which helps CHs estimate the density of the vehicles approaching the intersection. CHs transmit this density information to infrastructure nodes which can be used to support efficiency applications, such as, adaptive traffic lights. Souza et al. [SNG10] propose an algorithm to enhance the cluster lifetime which uses so-called Aggregate Local Mobility

(ALM) to improve cluster stability. This approach delays the re-organization of the clusters by delaying the status changes of CMs and CHs based on the ALM. Kayış et al. [KA07] propose a clustering algorithm that uses the speed and location of the vehicles to form the clusters. This approach uses multiple access control techniques and relays concepts to improve V2X communication efficiency. In this approach, CH is responsible to regulate the channel access of its CMs using the Code Division Multiple Access (CDMA) scheme. Inter-cluster communication is regulated using MultiCode Sense CDMA (MCS-CDMA) while gateway nodes are used to establish inter-cluster communication. Zhang et al. [ZSC06] propose a multi-channel clustering protocol to improve throughput and real-time delivery of the V2X communication. This approach uses different MAC channels for safety-critical and non-safety-critical message exchange. In this approach, CH is responsible to control the channel access for CMs and also, collection and delivery of safety messages within the cluster. Authors of [ZGZ+19] propose a general-purpose passive clustering algorithm to improve the stability of the vehicle clusters. This approach proposes a multi-hop clustering approach in which the vehicle and its most stable neighbor node within N hops become part of the same cluster. It uses WAVE architecture for intra-cluster communication and cellular network for communication with infrastructure.

Apart from these approaches, there are more efforts by the research community with focus on improving various aspects of VANETs, such as data dissemination, routing, channel access [GAS09, GWG07, QSW+18, AOW12, TBL10]. This shows that clustering in VANETs can be designed to cater to the various requirements of the desired application.

2.5 V2X Pilot Deployments

V2X technology provides much promise to enhance traffic safety and efficiency. However, technology apart, the actual traffic safety and efficiency processes are deployed and maintained by the city and highway authorities. With this perspective, V2X pilot project deployments have been undertaken with the following objectives:

- Bring different stakeholders, such as technology developers, traffic management authorities, public transport operators, together
- Evaluate the technology and its applications in different geographical locations with varying characteristics

- Understand the limitations of current traffic management infrastructure
- Develop the technology further with the inputs from real-life deployments

We describe a few selected projects here which can highlight the current deployment activities of V2X technologies.

2.5.1 North America

- New York City, USA - In this project, various V2X applications, based on DSRC, are deployed in three different areas [NYC15]. The deployment includes total of 8000 vehicles from various categories (public transport, cabs, etc.) and 300 RSUs. The deployed applications include V2V (Forward Crash Warning, Emergency Electronic Brake Lights, Vehicle Turning Right in Front of Bus Warning, etc.), V2I (Emergency Communication and Evacuation, Red Light Violation, Speed compliance, etc.), V2P (Pedestrian in Signalized Crosswalk, Mobile Accessible Pedestrian Signal Crosswalk).
- Tampa, USA - This project aims to improve public transport bus transit times, reduce congestion and collisions, and enhance pedestrian safety [USD18]. It deploys DSRC devices among 500 pedestrians, 10 transit buses, and 1600 passenger vehicles. The V2X applications include V2V (Forward Collision Warning, Emergency Electronic Brake Lights, etc.), V2I (Wrong Way Entry, End of Road Notification, Pedestrian Collision Warning, Pedestrian in Signalized Crosswalk, etc.).

2.5.2 Europe

- Kassel, Germany - City of Kassel aims to improve transit times of public transport vehicles (buses and trams) and emergency vehicles by connecting them with the traffic lights [Kas17]. In this project, public transport vehicles, ambulances, and traffic lights are equipped with C-ITS devices. The vehicles can communicate with the traffic light controller to request priority at the intersection. Also, emergency vehicles may communicate with public transport vehicles in order to notify their drivers that the emergency vehicle is on the way.
- Project Compass4D - This project involved the deployment of three C-ITS services in 7 cities across different countries in Europe [IRU15]. The services include Red Light Violation Warning, Road Hazard Warning, and Energy

Efficient Intersection (EEI). EEI further involved two applications. One of the applications is to have signal priority for public transport, emergency, and heavy vehicles. Another application from EEI involved Green Light Optimum Speed Advisory (GLOSA) which calculates optimum speed to reach the intersection to catch the traffic light is green thereby reducing the wait time and emissions at the intersection.

2.5.3 Asia

- Wuxi, China - City of Wuxi has deployed C-V2X-enabled devices for infrastructure and vehicles [TAA18]. This project enables multiple connected car applications, such as Emergency Electronic Brake Light, Vulnerable Road User Collision Warning, status and duration of traffic lights, etc.
- Ahmedabad, India - City of Ahmedabad has deployed a pilot V2X project that involves road-side displays for ambulance vehicle approach warning [Tod19]. Ambulance and the display unit communicate with each other through V2X technology. The display then shows the recommendation to move to another lane so that the ambulance can pass through.

2.6 Discussion

V2X domain is a broad area spanning multiple use cases, road entities, traffic situations, supporting technologies, and geographical locations. Although technology standardization efforts have made big strides in harmonizing these broad areas, V2X cannot be considered as one monolithic system due to its varied nature. Above all, most of the V2X communication is context-dependent and has limited geographical scope. Also, as the vehicles and VRUs are mobile, their context changes with the geographical location. This implies that the different components of V2X such as V2V, V2P, may exist as separate, interactive, and codependent subsystems as part of a larger V2X system.

The design of an effective V2X system requires that characteristics of its subsystems be studied, interaction and codependency among subsystems be evaluated and appropriate optimization mechanisms be developed. As the V2P subsystem has started to emerge in the last few years, it too needs to go through these development cycles. In the next chapter, the V2P subsystem is discussed in detail outlining its characteristics, current state-of-the-art, and its challenges.

3 Vehicle-to-Pedestrian Communication Systems

As discussed in section 1.1, worldwide crashes involving the VRUs are still high despite equipping the vehicles with ADAS. V2X systems have been designed to extend the ADAS capabilities further, however, the V2P component of the V2X system is still in a nascent stage. This section provides an overview of current V2P efforts and discusses its challenges.

3.1 V2P Introduction

V2X communication capabilities can be extended to include VRUs in order to enhance the crash-prevention system of the vehicles further. For this purpose, VRUs and vehicles require to exchange information among themselves. Vehicles can be equipped with On-Board Units (OBUs) that broadcast their current status, such as location, speed, direction, etc. However, to establish such communication with vehicles, VRUs too need to carry a device that is equipped with Inertial Motion Sensors (IMUs), positioning capabilities, such as Global Positioning System (GPS), communication technology, display, and speaker for audio-visual warnings, and enough computing power. As smartphones are ubiquitous and already capable of providing such functionality, smartphones are the strongest candidate to perform the role of VRU devices in a V2P system. Although current off-the-shelf smartphones are not equipped with the required communication technology (IEEE 802.11p), efforts have already shown that it is possible [WMY+14]. Such Vehicle-to-VRU communication may be called as Vehicle-to-Pedestrian (V2P) communication in which V2P encompasses the exchange of information between all types of VRUs (pedestrians, bicyclists, and MTWs) and vehicles.

V2P communication can be used for safety applications as well as non-safety applications. Examples of safety applications are Collision Warning, Pedestrian in the Crosswalk, etc. Due to its importance, multiple efforts have been made to develop prototypes of collision warning safety application [WMY+14, SNH08,

DSCP14, FSYN15, LFYZ14, TSKF$^+$15, BSN14, HC17]. Examples of non-safety applications include traffic advisory and road incident notifications. [TIM17] and [TM13] are examples of such non-safety applications.

The safety application efforts, that are mentioned above, design a V2P collision warning system that involves vehicles and VRUs exchanging safety messages. In presence of a large number of VRUs, the large number of VRU-generated safety messages would cause network congestion and affect safety-critical V2V communication. There have been multiple efforts to reduce the network congestion in V2V network [GAR17, ELPM16, WJS17, MOL15, ETS18b]. However, these efforts focus on reducing V2V network congestion in the case of a large number of vehicles. A mechanism to reduce the congestion caused by VRUs is currently absent. Also, when a vehicle and a pedestrian are on the verge of collision, their communication requires higher priority compared to the other non-critical communication. Such provision of Quality of Service for crucial communication is currently not possible. In this section, we look at various aspects of V2P communication and discuss the associated challenges.

3.2 Fundamentals of V2P Crash Prevention System

Any crash prevention system requires that the potential hazards be detected in time and appropriate action be taken. We discuss such aspects of V2P communication, for safety application, in this section.

3.2.1 Contact Duration

Contact duration for V2P is defined as *the length of time during which a vehicle and a VRU can communicate with each other* [SKS17]. The factors impacting the contact duration are the distance between the nodes, node speed, communication range, and movement direction. In a typical urban intersection scenario, there are three types of possible node movements as follows: node moving in the same direction, nodes moving in the opposite direction, and nodes traveling in the orthogonal direction. Figure 3.1 illustrates these possible scenarios.

1. Nodes traveling in same direction - In this scenario, the vehicle and VRU travel in the same direction, and the vehicle is approaching the VRU. Contact duration for this scenario is the length of time between the time since the beginning of their communication and the time when the vehicle passes (or crashes into) the VRU. If the communication range of VRU node can be

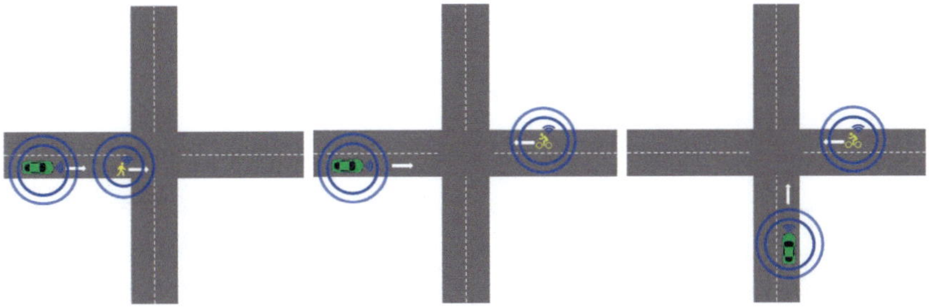

(a) Same direction (b) Opposite direction (c) Orthogonal Direction

Figure 3.1: Contact duration scenarios

given by r_p, VRU speed v_p, and vehicle speed v_v then the contact duration can be calculated using Equation 3.1 [SKS17].

$$t_c = \frac{r_p}{v_v - v_p} \tag{3.1}$$

2. Nodes traveling in opposite direction: In this scenario, the vehicle and VRU travel towards each other in the opposite direction. Contact duration for this scenario is the length of time between the time since the beginning of their communication and the time when both the nodes cross (or crash into) each other. If the communication range of VRU node can be given by r_p, VRU speed v_p, and vehicle speed v_v then the contact duration can be calculated using Equation 3.2.

$$t_c = \frac{r_p}{v_v + v_p} \tag{3.2}$$

3. Nodes traveling in orthogonal direction - In this scenario, the vehicle and the VRU travel in the orthogonal direction towards the intersection. If the communication range of VRU node can be given by r_p, VRU speed v_p, and vehicle speed v_v then the contact duration can be calculated using Equation 3.3 [SKS17].

$$t_c = \frac{\sqrt{2} \cdot r_p}{v_v + v_p} \tag{3.3}$$

Contact duration directly affects the number of safety messages that can be exchanged between the vehicle and the VRU.

3.2.2 Phases of Crash Prevention

V2P crash prevention system consists of three phases: *detection*, *tracking and prediction*, and *action* [SKS17].

- *detection*: This phase involves the first contact between the vehicle and the VRU. In this phase, the vehicle and VRU both receive the safety messages from each other for the first time and become aware of each other's presence. It may happen that the vehicle and VRU transmit the safety messages with different transmission power. For example, vehicle OBU may transmit with higher power than the pedestrian device due to more battery capacity and hence its messages may reach out farther. In this case, the pedestrian device may become aware of the vehicle before the vehicle becomes aware of the pedestrian. In order to allow sufficient time for the *tracking and prediction* phase, first contact or *detection* should occur as early as possible.

- *tracking and prediction*: After the initial contact, the vehicle and VRU may start tracking each other. As nodes transmit the safety messages periodically, information from the received messages, such as location, speed, direction, etc., may be used to calculate the prediction of the trajectory. This calculation then may be compared with its own trajectory prediction in order to calculate the probability of the collision. The prediction and the probability of collision need to be reliable in order to make the decision about *action* that needs to be taken. This requires that the information needed for prediction, such as speed, location, direction, be as accurate as possible. Besides, prediction also requires sufficient information to predict reliably. In other terms, the node must receive a sufficient number of safety messages to make a reliable prediction of trajectories of other nodes.

- *action*: After *tracking and prediction* phase predicts that the two nodes are on the verge of collision, appropriate action must be taken to avoid the crash. For the vehicle, such action would be to initiate emergency braking or an evasive maneuver. In the case of VRU, this may be an audio-visual warning. However, the notification mechanism and notification recipient, especially on the VRU side, depend on the design of the system. Also, if one peer node predicts the crash before the other peer node, it may inform the other node by sending a highest priority warning message [SKS17].

SAE J2945/9 specification requires that vehicles must receive the crash warning message at least 8 s before the crash [RCL+16]. This crash warning message corresponds to the *action* phase and needs to be observed by the vehicle manufacturers.

3.2.3 V2P System Architecture

V2P system for collision avoidance involves VRUs and vehicles exchanging safety messages periodically among each other. This exchange can happen in the following ways:

- Direct communication: In this mode, vehicle OBU and VRU device exchange safety messages *directly* i.e. in an ad-hoc manner without any infrastructure using technologies, such as IEEE 802.11p.
- Indirect Communication: In this mode, vehicle OBU and VRU device exchange safety messages *indirectly* i.e. through infrastructure mode using technologies, such as cellular communication.
- Hybrid communication: In this mode, multiple communication technologies may be used to exchange safety messages. For example, VRU and vehicle devices may use different technologies to communicate based on the scenario latency requirement. Another example is that an infrastructure node may use multiple technologies to receive and forward safety messages among the devices equipped with different technologies.

Based on the design, the V2P system may comprise of following 4 components [SS19b]:

1. VRU device
2. Vehicle OBU
3. Information processing unit
4. Infrastructure

In the case of a *direct* system, it comprises of a VRU device and a vehicle OBU which exchange the safety messages directly. In such a system, both devices are responsible for carrying out all three phases of crash prevention. In the case of an *indirect* system, the VRU device and vehicle OBU exchange safety messages through the infrastructure, such as cellular network, or RSU. In such system, the *Information processing unit* may reside as a separate entity which carries out the *detection* and *tracking and prediction* tasks. Based on its recommendation, VRU

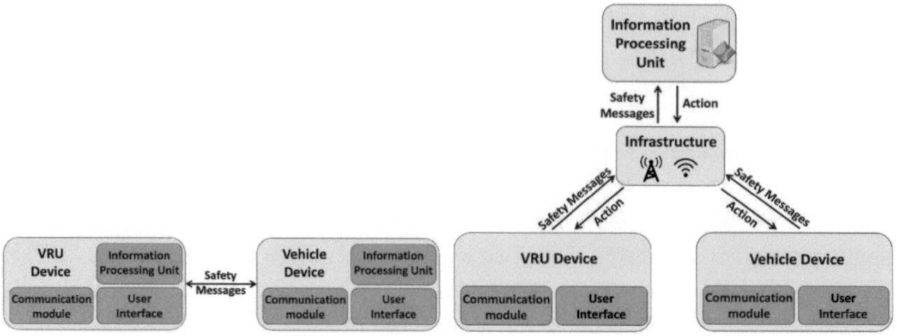

(a) Direct Communication (b) Indirect Communication

Figure 3.2: Example V2P system architecture [SS19b]

and vehicle device may carry out the appropriate *action*. Figure 3.2 illustrates example architectures of the V2P *direct* and *indirect* communication system.

3.3 VRU Characteristics

VRUs differ from vehicles in mobility, traveling patterns, etc. Moreover, VRUs themselves differ from each other in such characteristics. This makes it necessary to understand various VRU characteristics in order to design an effective V2P system and communication protocol. In this section, we discuss characteristics of VRUs that are important for V2P system design.

- Mobility - Different VRU groups have different mobility patterns. Pedestrians typically walk at a speed of 1.4 m/s (5 km/h). Cyclists' typical traveling speed is 4.2 m/s (15 km/h) while that of MTWs is 14 m/s (50 km/h). Also, pedestrians walk on the footpath while cyclists and MTWs travel on the road or dedicated paths. Such differing mobility characteristics have an impact on the contact duration and the time available for various phases of crash prevention. Also, pedestrian walking patterns are more unpredictable than the other two VRU groups which make the prediction harder.
- Topology - Due to their differing mobility characteristics, the topology of V2P communication is also affected. Due to the slow walking speed, communication between pedestrians and vehicles is mainly affected by vehicle mobility. In the case of cyclists and MTWs, topology changes would be more frequent due to their higher speed.
- Energy constraints - VRU device may differ for different VRU categories.

For pedestrians and cyclists, a smartphone is the logical choice as a VRU device. For MTWs, this may be an OBU specially designed for MTW needs. A system that uses a smartphone as a VRU device may need to be more energy-efficient than a system using a dedicated OBU.

- Node density - VRUs can be present in large numbers in a given geographical location. Locations, such as stadiums, train stations, can have a disproportionately large number of VRU presence and hence V2P systems may need performance optimization for such cases.

3.4 V2P System Requirements

3.4.1 V2P System Design

3.4.1.1 Overview

VRUs can be distinguished by their properties, such as speed, traffic rules, mobility, and travel patterns. For example, MTWs and bicyclists travel faster than pedestrians. Another example is pedestrians can be present in significantly larger numbers than the other two VRU types. These characteristics can affect not only V2P but also V2V and V2I communications and hence, are vital in designing an efficient V2P system [SS19b]. V2P system developers, standardization bodies, and traffic management authorities need to consider such characteristics while they design the system.

3.4.1.2 Related Work

Multiple efforts have been made to design a V2P system which use various communication technologies [ATG$^+$15, WMY$^+$14, SNH08, DSCP14, FSYN15, LFYZ14, TSKF$^+$15, BSN14, HC17, Gro09]. Authors of [WMY$^+$14] propose a DSRC-based V2P system for collision avoidance. This system uses a DSRC-equipped smartphone as a VRU device that exchanges safety messages with a vehicle OBU. Smartphone and OBU process these messages to calculate the crash probability and notify the user (VRU or driver) if required.

Anaya et al. [AMS$^+$14] propose V2ProVu, a WiFi-based V2P system, for alerting pedestrians about imminent collision. Pedestrian's smartphone receives the safety messages from vehicle OBU which are then processed to calculate the probability of collision. In this system, the pedestrian device does not send any safety messages and is solely responsible for calculating the crash probability. WifiHonk

V2P System	Application Type	Technology
Nguyen et al. [NMKF20]	safety	Cellular/ Multi-Edge Computing
Wu et al. [WMY+14]	safety	802.11p
WiSafe [HC17]	safety	Wi-Fi
WiFiHonk [DSCP14]	safety	Wi-Fi
V2ProVu [AMS+14]	safety	Wi-Fi
Lee and Kim [LK16]	safety	802.11p
Merdrignac et al. [MSN17]	safety	Wi-Fi
Bagheri et al. [BSN14]	safety	Cellular
Zadeh et al. [ZGE18]	safety	Cellular
Nakanishi et al. [NYF+10]	safety	Wi-Fi
V2PSense [LSH+18]	safety	Cellular
LP^3S [LBKW13]	safety	802.15.4
pSafety [LCC+16]	safety	Cellular
Audi [Eng13]	safety	Wi-Fi
General Motors [GMW12]	safety	Wi-Fi
Artail et al.[AKY17]	safety	802.11p, Cellular
POFS [LPM+15]	safety	Cellular, WiFi
Sugimoto et al. [SNH08]	safety	Cellular, WiFi
Fujikami et al. [FSYN15]	safety	Wi-Fi
Hussein et al. [HGAOM16]	safety	Wi-Fi
Tahmasbi-Sarvestani et al. [TSKF+15]	safety	802.11p
Thielen et al. [TLH+12]	safety	Wi-Fi, 802.11p
Ko-TAG [SKCB12]	safety	localization
Nagai et al. [NND12]	safety	700MHz ITS
RedEye [HCT+16]	safety	Wi-Fi
C-AEB [KOV+14]	safety	802.11p
Liu et al. [LLM+16]	safety	Wi-Fi
Hernandez-Jayo et al. [HJPI15]	safety	Cellular, 802.11p
MotoWarn [ATG+15]	safety	Bluetooth, 802.11p
David and Flach [KA10]	safety	Cellular, WiFi
TIMON [TIM17]	convenience	Cellular, 802.11p
Lu et al. [LFYZ14]	convenience	802.11p

Table 3.1: Summary of current efforts for V2P systems

is a V2P system proposed by Dhondge et al. [DSCP14] that uses WiFi to enable V2P communication. It uses WiFi beacons stuffed with context information as safety messages. Field experiments of this system show that V2P communication can be achieved within the range of 50 m and with a speed up to 70 mph (112 km/h). Authors of [SNH08] propose a hybrid system for V2P communication that uses cellular and WiFi communication for the exchange of safety messages. VRU device (cellular phone) and vehicle OBUs send information to a dedicated server which estimates the collision risk. If the risk is deemed high, it instructs the VRU and vehicle devices to establish direct communication using WiFi. Authors of [HJPI15] propose CS4VRU, a collision warning system, for cyclists. Cyclists and vehicles exchange safety messages through a hybrid system consisting of IEEE 802.11p and cellular communication. An RSU is responsible for forwarding the messages over heterogeneous communication technologies. The cyclist's smartphone is connected to a LED-equipped helmet via Bluetooth. The helmet flashes red LEDs upon the indication of potential collision coming from the smartphone. MotoWarn system [ATG+15] for MTWs and cyclists uses DSRC and Bluetooth respectively for V2P communication. MTW uses DSRC-equipped OBU in order to exchange safety messages with vehicle OBU while the cyclist uses iBeacon technology to convey its location to vehicle OBU. In this case, vehicle OBU is equipped with multiple technologies, such as DSRC, WiFi, and Bluetooth. Vehicle OBU also connects to the driver's smartphone using WiFi. The smartphone acts a *Information processing unit* in this case. Ko-TAG project [SKCB12] develops a localization technology using 5.768 GHz as a center frequency. This approach equips pedestrians with a transponder tag (VRU device) operating at the center frequency. Vehicle OBU communicates with this VRU device to measure the flight round-trip which is then used to determine the distance between the vehicle and pedestrian. Table 3.1 shows summary of V2P system design efforts.

Apart from such efforts, there are also various pilot projects, such as XCYCLE, VRUITS, PROSPECT, InDev, that have been undertaken to understand different aspects of targeted VRUs groups [XCY18, VRU16, Pro17, InD18].

3.4.1.3 Analysis

Although there have been numerous efforts to prototype V2P systems, they lack a systematic and holistic approach in the design process. They use various communication technologies, different VRU devices, and varying warning mechanisms. Also, there is a lack of understanding of the required periodicity of safety messages, the impact of the periodicity of the safety messages on crash prevention phases, and the required communication range. This also leads to a lack of understanding of the VRU devices' roles in V2P communication.

This demands that V2P communication be analyzed further from a holistic perspective. With this understanding, a systematic approach to design the V2P system can be developed.

3.4.2 Network Congestion Control

3.4.2.1 Overview

DSRC and C-ITS-based V2V communication leads to a decentralized broadcast network where channel access is controlled using CSMA/CA mechanism. This makes the network prone to congestion when a large number of vehicles are present [RCB+16]. This problem worsens when V2P communication is enabled and when VRUs are present in large numbers. The VRU-generated safety messages can congest the network further [RCL+16, SKS17].

Due to the single-hop broadcast nature of the V2X communication, the congestion is seen mainly at the MAC layer channels. Due to this, the network load is measured as a load on the MAC layer channel. The channel load is measured using the metric Channel Busy Ratio (CBR), also referred to as Channel Busy Percentage (CBP). CBR values can be used to quantify the network load at three different levels: RELAXED (CBR $<$ 30%), ACTIVE (30% $<$ CBR $<$60%), and RESTRICTIVE (60% $\leq$ CBR) as defined by ETSI [ETS18b].

3.4.2.2 Related Work

Multiple congestion control mechanisms have been proposed that focus on V2V network congestion issue [MLGN17, AS13, BLKP13, BKR13, DGD12, ETS18b, JLJ18, ELPM16]. The congestion control approaches can be further divided into six categories which are discussed below:

1. Adaptive Beaconing Rate Approach: In this approach, vehicles vary the safety messages periodicity (beaconing rate) based on the channel load. Under normal conditions, vehicles are required to broadcast beacons at 10 Hz. However, when the channel load increases, the adaptive beaconing policy requires that the vehicle transmits the beacons at the rate $<$ 10 Hz. This reduces the number of packets being transmitted thereby reducing the network load. FABRIC congestion control algorithm [ELPM16] uses a concept called as *weighted utility function* of a vehicle. It employs the Networks Utility Maximization (NUM) principle to determine the weights of the utility function of each vehicle. Based on the assigned weights, the beaconing rate of the vehicle is then determined. Authors of [KBR11] propose an adaptive beacon rate control algorithm called LIMERIC. This algorithm uses the

measured values of the network capacity currently in use which can be measured by every user and also, the estimated number of surrounding vehicles. This information is then used to make the decision about the beaconing rate. Huang et al [HFSK11] propose an adaptive beaconing algorithm that embeds the channel state information into the safety messages. Each vehicle tracks the neighboring vehicles and their channel state information based on the received safety messages. Based on the received state information, it adjusts the beaconing rate.

2. Dynamic Priorities Approach: As we discussed in 2.3.1.2, IEEE 802.11p uses EDCA priorities in order to prioritize the channel access. This approach involves varying the EDCA priorities, Contention Window (CW), or AIFS of safety messages to control the network congestion and improve channel access to high-priority messages. Taherkhani et al. [TP16b] propose an algorithm that assigns priorities to safety messages based on its application and also provides dynamic scheduling of such messages. Authors of [SCZ+13] propose a distributed and multi-priority congestion control algorithm. This algorithm varies the contention window size dynamically based on the congestion status. Also, beacons with different priorities are assigned the appropriate CW size. Hsu et al. [HHT11] propose an Adaptive Offset Slot (AOS) mechanism that adjusts the minimum size of CW based on the number of vehicles. Each vehicle estimates the number of neighboring vehicles based on the received messages and then picks appropriate CW_{min} size in order to maximize the channel access.

3. Transmission Power Control Approach: The communication range of the vehicle directly depends on the transmission power used for safety messages. Reduced transmission power leads to the delivery of safety messages only in the vicinity thereby mitigating the congestion. Authors of [WJS17] propose a dynamic transmission power algorithm that toggles the power between high and low levels. The power levels can be changed dynamically based on the network congestion level. This approach guarantees the delivery of messages to vehicles in close proximity. Joseph et al. [JLJ18] propose a speed-based adaptive transmission power algorithm (SBAPC) that adjusts the power levels according to the vehicle's speed. As speed has a direct correlation with the Time-To-Collision (TTC), higher speeds lead to higher TTC and hence safety messages need to reach farther. For these reasons, this approach employs higher transmission power for higher speeds. Goudarzi et al. [GAR17]

propose a game-theory based approach to determine the transmission power on the beacons. This approach employs the Nash Equilibrium method to find the equilibrium in the transmission power being used and the Channel Busy Ratio (CBR) which is an indication of network load. Appropriate power is then selected to maintain the desired CBR.

4. CSMA/CA MAC Approach: As we discussed in 2.3.1.2, CSMA/CA is a channel access control mechanism being used by IEEE 802.11p. However, this mechanism is designed to suit a broader range of wireless devices. By modifying the channel access mechanism to suit V2X needs, the capacity of the channel can be increased, in turn, reducing the network congestion. Stanica et al. [SCB12] propose a new channel sensing mechanism, called Safety Range CSMA (SR-CSMA). This approach considers the location of the transmitter, in addition to the channel state, to determine the appropriate transmission opportunity. Location is also considered while starting the back-off timer. This approach improves the channel access for the vehicles that are located in the vicinity of each other.

5. Adaptive Data Rate Approach: This approach involves adjusting the data rate to different levels of channel load. Math et al. [MOL15] propose an extension of the ETSI-formulated Decentralized Congestion Control (DCC) algorithm that varies the data rate based on the channel load (CBR). The approach measures the CBR periodically and then adjusts the data rate accordingly if pre-defined thresholds of CBR have been changed. Higher CBR leads to higher data rates allowing more messages to be transmitted which effectively lowers the CBR.

6. Hybrid Approach: Hybrid approach combines two or more approaches described above to achieve congestion control. C-ITS architecture uses the Decentralized Congestion Control (DCC) mechanism as standardized by ETSI [ETS18b]. DCC mechanism employs Transmit Power Control (TPC), Transmit Rate Control (TRC), and Transmit Datarate Control (TDC) approaches for congestion control. DCC also uses reactive and adaptive approaches which are executed at different levels of congestion. WAVE architecture employs a congestion control mechanism as described by SAE J2945/1 standard [RKG18]. This algorithm employs an adaptive congestion control mechanism that uses varying beaconing rates and varying transmission power. Every vehicle node counts the neighboring vehicle nodes within 100 m periodically

and adjusts the congestion controlling parameters. Taherkhani et al. [TP16a] propose an infrastructure-based congestion control strategy where RSUs are responsible for detecting the congestion and controlling network parameters to mitigate it. In this approach, RSUs classify the messages based on their contents and decide the appropriate transmission range, contention window, and AIFS size for the message category. Bansal et al. [BLKP13] propose EMBARC algorithm which combines the LIMERIC algorithm (described in *Adaptive Beaconing Rate Approach* section) with a dynamic scheduling mechanism. This approach uses vehicle movement to make decisions about the scheduling. This leads to a higher beaconing rate for faster moving vehicles. Authors of [KHS+16] propose a Random Transmit Power Control (RTPC) algorithm that employs transmit power control along with beaconing rate control to mitigate the congestion. As RTPC employs randomized transmit power, it eliminates the packet collisions happening due to hidden node problems. Similarly, there are efforts by research community [MLGN17, DGD12, TJHD13] that use combination of two or more congestion control approaches.

3.4.2.3 Analysis

All of the congestion control approaches discussed above are specially designed with V2V communication as a primary focus. Also, our discussion in section 2.4.2.2 shows that although VANET clustering approaches may provide mechanisms to improve channel access and reduce congestion, they are also primarily focused on V2V communication. These approaches require that all the nodes involved in communication must take action to control the congestion (in a reactive or adaptive) manner. Also, the current approaches are unable to differentiate among different types of nodes (vehicles, VRUs, etc.). When V2P communication is enabled, it is expected that it would cause additional network load. However, in a scenario with high VRU density, channel load can quickly increase due to a large number of VRU-generated safety messages. Due to such high channel load, safety-critical V2V communication can get affected. With current congestion approaches, vehicles would be required to adapt to the high channel load. This is especially a concern because, with moderate vehicle density, vehicles can be highly mobile and may require a longer range of communication or maximum beaconing rate to maintain reliable safety-critical communication. Also, vehicles may use different applications based on their mobility even in the same geographical area. For example, vehicles in one direction may be traveling slow and hence, maybe using a brake-assist application. Vehicles in the other direction may be traveling fast due to the lower number of vehicles and hence, maybe using lane-change-assist application. Under such conditions, V2V awareness requirements are different. If vehicles are required to take measures to mitigate the network congestion, their awareness requirements may not be satisfied [SMS+11]. Also, although the periodicity of the VRU-generated safety messages can be determined based on the context of the VRU, the network can still be congested in presence of a high number of VRUs [RCL+16]. This demands that a network congestion control mechanism, which considers VRUs separately, is required. This mechanism should be able to control the channel load caused by VRU-generated safety messages without affecting the V2V communication. It should also be able to support reliable V2P communication.

3.4.3 Quality of Service for Crucial V2P Communication

3.4.3.1 Overview

Similar to V2V broadcast nature, V2P is also expected to be broadcast in nature. Based on the received safety messages, the phases of crash prevention, as described in Section 3.2.2, track all surrounding nodes and make predictions of collision continuously. After a vehicle or a VRU determines that they are on the verge of collision, their communication becomes crucial and safety-critical. Their communication requires the highest priority among all the other communication (higher than even the other safety-critical communication). This requires the highest QoS among all the other V2X communication. Despite the broadcast nature of V2X, the communication between this particular pair has peer-to-peer nature where they must continue to *track and predict* each others' movements till the necessary *action* is taken. In order to achieve this, they need to exchange safety messages with the highest periodicity (10 Hz) reliably.

3.4.3.2 Related Work

As V2X communication by itself is safety-critical, there have been multiple efforts to improve the QoS of the V2X protocols. These efforts are discussed in this section.

Boulila et al. [BHLS18] propose a MAC protocol called QCH-MAC (Qos-aware Centralized Hybrid MAC). This approach combines TDMA and EDCA access mechanisms to improve the QoS. It requires infrastructure nodes, such as RSU, to assign time slots to provide transmission opportunities. Chang et al. [CYD16] propose a channel access mechanism called as *Earliest Deadline First based CSMA* or EDF-CSMA. This approach forms clusters of vehicles where Cluster Head (CH) first gathers requirements of QoS from Cluster Members (CM). CH is then responsible to maintain the QoS allocations (channel access control) for the CMs. Authors of [ZTL18] propose improvements to the existing EDCA mechanism of 802.11p to improve guarantee timely delivery and also, network capacity. This improvement involves Dynamic Queue Management mechanism which estimates the delays in different Access Category (AC) queues to determine the queue with lowest transmission delays. The improvement also involves a dynamic back-off mechanism which involves changes in the size of CW_{min}. Ouni et al. [OBZ17] propose a channel access mechanism that guarantees delivery of critical messages within a specified time-window. This approach modifies EDCA to change the CW for various ACs in order to provide predictable transmission delay. Nasrallah et al. [NAM17] propose improvements in assigning AIFS values for channel access. In this approach, higher priority AC always gets access to the channel despite the leftover CW value for lower priority AC. It also provides an adaptive mechanism to determine AIFS values for each AC. Authors of [KKJ18] propose a protocol to improve QoS in the environment with frequent link disconnection. This protocol provides multi-hop, infrastructure-based relay communication for the vehicular nodes with weak links to infrastructure. Wang et al. [WLZF11] propose QoS Variable CCH Interval (Q-VCI) MAC protocol which can support QoS in multi data-rates and multi-channel operation. This mechanism uses variable channel operation interval instead of the fixed interval of 50 ms as shown in Figure 2.5. This approach also uses a varying minimum CW for different priorities. Authors of [APB12] propose Multichannel QoS Cognitive MAC (MQOG) mechanism that selects the optimum channel for transmission for QoS. The optimum channel is determined based on the measurements of noise and interference on all channels. This approach defines special control frames that every vehicle needs to transmit, on the CCH, to its neighboring vehicles. These control frames include information regarding the channel condition parameters, channel selection, and notification to switch channels.

3.4.3.3 Analysis

All of the QoS improvement efforts have focused on improving channel access and channel capacity. These efforts apply the improvement to the broader V2V communication and do not focus on any particular transmitter-receiver pair which requires the highest priority and unfettered channel access. A pair of vehicles and VRU, that are on the verge of collision, can be considered as such a transmitter-receiver pair.

As discussed in section 3.2.2, the V2P collision avoidance system operates in three phases viz. *detection*, *tracking and prediction*, and *action*. During the *tracking and prediction* phase, the vehicle-VRU pair may determine the probability of collision. The probability may be low at first and may continue to increase to a point when it is highly probable that the pair will crash into each other. The prediction with high probability is also necessary to avoid false-positive warnings during the *action* phase. However, to predict with high probability, the pair must be able to communicate with each other reliably during the prediction phase. Such a differentiation of reliability for the crucial communication is currently not supported in the existing QoS mechanisms.

Also, there is currently no way for surrounding nodes to know about the existence of such crucial communication. As V2X is based on cooperative safety, it is imperative that the surrounding nodes know about the crucial communication taking place. A mechanism is required which would provide the highest priority for the crucial communication and also, inform the surrounding nodes about the presence of such crucial communication.

3.5 Evaluation of V2P Communication

It is expected that VRU-generated safety messages would increase the load on the V2V network. However, it is still necessary to understand the various aspects of VRU-generated safety messages. This raises the following questions:

- What are the effects of an increasing number of VRUs on the V2V channel load?
- What are the effects of the different periodicity of VRU-generated safety messages on V2V channel load?
- How does the network load affect the communication between a pair of vehicle and pedestrian that are on the verge of collision?

We devised a scenario that helps us find the answers to these questions. We discuss the details of our evaluation below.

3.5.1 Scenario

In order to evaluate the network congestion caused by VRUs, we require a scenario that allows us to create dense pedestrian traffic. We also require vehicles and pedestrians that are on the verge of collision to evaluate the performance of crucial communication. Vehicle and pedestrian crashes happen predominantly under the following two pre-crash scenarios:

- Scenario 1 - Pedestrian crossing the road in front of the vehicle that is traveling in a straight line. 88% of the estimated fatalities happened under this pre-crash scenario in the USA between 2005 - 2009 [CFM+13].
- Scenario 2 - Pedestrian walking parallel to the vehicle's direction of travel. 12% of the estimated fatalities happened under this pre-crash scenario in the USA between 2005 - 2009 [CFM+13].

These pre-crash scenarios help define the mobility model of the vehicles and pedestrians for simulating the crashes.

We designed a typical urban intersection scenario with the traffic of vehicles and pedestrians as well as two pairs vehicles and pedestrians that simulate the pre-crash scenarios mentioned above. Figure 3.3 shows the layout of the intersection. The

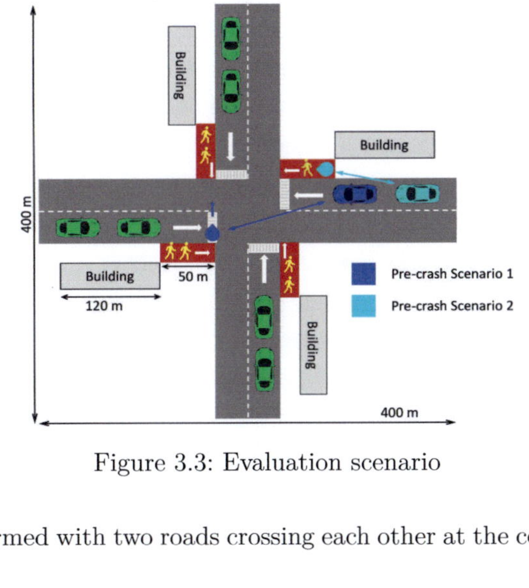

Figure 3.3: Evaluation scenario

intersection is formed with two roads crossing each other at the center. Both roads are 400 m long and have 2 dedicated lanes in each direction. Vehicle flow (traffic) at the intersection is controlled using traffic lights. All the vehicles enter at one end of the road and travel in a straight direction to the other end of the road without making any left or right turns. The vehicles travel at a typical urban speed of 50 km/h.

There are four footpaths in the scenario that are of 50 m length each. All of the pedestrians are inserted at the far end of the footpath. They travel straight towards the intersection. We vary the number of pedestrians to change the network load generated by VRU safety messages. We simulate two pre-crash scenarios described above using two pairs of vehicles and pedestrians as shown in Figure 3.3.

We use WAVE architecture for our evaluation scenario. We assume that the vehicles transmit safety messages with a fixed 10 Hz periodicity. Pedestrian beaconing periodicity can be varied based on the pedestrian's current context (standing, walking, etc.) [RCL+16]. We vary the periodicity of the pedestrian beacons which is chosen from four different values of 1, 2, 5, and 10 Hz.

3.5.2 Simulation Environment

We use the combination of OMNeT++, Veins, and Simulation of Urban MObility (SUMO) [omn, SGD11, KEBB12] to simulate the WAVE architecture based V2V and V2P network. Vehicles are inserted every 1 second at one end of all lanes in four different directions. We vary the number of pedestrians between 0-120. We achieve this variation by varying the insertion interval of pedestrians in SUMO trace files. As pedestrian node insertion and pedestrian node exit from the simulation is not synchronized due to varying insertion intervals, number of pedestrians may vary by ±2 from the desired number. We set vehicle and pedestrian transmission power to 20 mW each which achieves a communication range of 400 m. We employ a realistic *TwoRayInterferenceModel* signal propagation model which provides us accurate estimation of signal path loss in urban environment [SGD11]. Table 3.2 provides further details of our simulation parameters.

The simulation warm-up period is set to 30 seconds so that the simulation reaches a state of busy urban intersection. The traffic light changes its green phase to yellow phase after 31 s. It changes its phase from yellow to red after 4 seconds. During this time, the traffic in the orthogonal direction gets a red traffic light for 35 seconds. The total length of simulation of 70 seconds in order to capture one cycle of a traffic light. Each simulation configuration is run three times with different seeds and the mean values of gathered data are used for the

Parameter	Value
Road length	400 m
No. of vehicles	100–120
Max. vehicle speed	13.89 m/s = 50 km/h
No. of pedestrians	0, 18 ± 2, 46 ± 2, 66 ± 2, 90 ± 2, 120 ± 2
Max. pedestrians speed	1.3 m/s
Vehicle transmission power	20 mW
Pedestrian transmission power	20 mW
Data rate	6 Mb/s
Vehicles beacon periodicity	10 Hz
Pedestrian beacon periodicity	1, 2, 5, 10 Hz
Beacon length	1024 bits

Table 3.2: Simulation parameters [SKS17].

evaluation.

3.5.3 Evaluation

3.5.3.1 Network load

To measure the impact of V2P communication on V2V, we measure network load (channel load) and Beacon packet Error Ratio (B_pER) as reported by every vehicular node. We use the metric Channel Busy Percentage (CBP) to measure the channel load as shown in Equation (3.4) [SKS17].

$$CBP = \frac{t_{\text{ChBusy}}}{t_{\text{CBP}}} \times 100\% \tag{3.4}$$

where:

t_{ChBusy} = channel busy time reported by each vehicle node during the measurement window

t_{CBP} = length of measurement window

As vehicles may be present for different duration of time during the simulation while data is being gathered, this duration becomes *lifetime* of the vehicle. This *lifetime* corresponds to the t_{CBP} for the vehicle. CBP is calculated only for vehicle nodes as we want to measure the impact of V2P on V2V. We then use the mean value of the CBP reported by all vehicle nodes. Figure 3.4 shows the CBP values with the varying number of pedestrians and varying pedestrian beacon periodicity. As the figure shows, CBP at the vehicle nodes increases with the number of pedestrians and pedestrian beaconing periodicity. As we discussed in section 3.4.2, the CBP levels are defined as: RELAXED (CBP $<$ 30%), ACTIVE (30% $<$ CBP $<$60%), and RESTRICTIVE (60% $\leq$ CBP). Vehicles are required to adapt the beaconing periodicity to keep the CBP as low as possible. For example, vehicles are required to lower the beaconing periodicity to keep the CBP going from RELAXED to ACTIVE level. Such reduction in periodicity may have adverse effects on safety-critical information dissemination among vehicles. Figure 3.4 shows that in presence of 90 and 120 pedestrians transmitting with 10 Hz periodicity, CBP changes from RELAXED to ACTIVE level. This would require vehicles to employ the congestion control mechanisms.

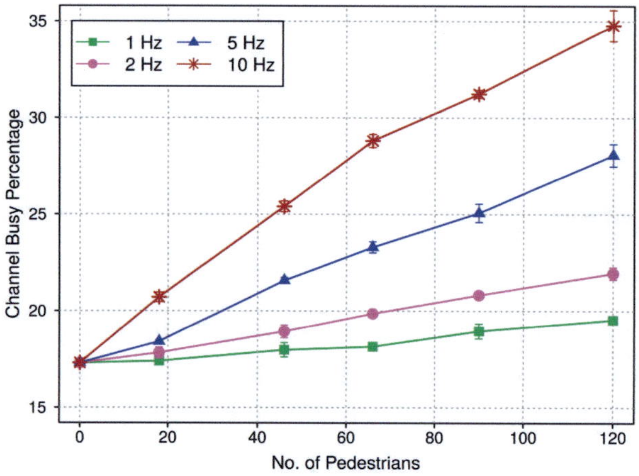

Figure 3.4: Channel Busy Percentage [SKS17]

We also studied the CBP distribution reported by individual vehicle nodes for the configurations with high number of pedestrians and high beaconing periodicity. Figure 3.5 shows the distribution of CBP for such configurations. The figure confirms that for 90 and 120 pedestrians with beaconing periodicity of 10 Hz, average CBP is correctly at the ACTIVE level. Figure 3.4 shows that for the configuration with 66 pedestrians with 10 Hz, and 120 pedestrians with 5 Hz, the average CBP is at RELAXED level. However,as per Figure 3.5, there are at least few vehicles nodes for which the CBP is at ACTIVE level. For 46 pedestrians with 10 Hz, and 90 pedestrians with 5 Hz configurations, there are at least few vehicular nodes with CBP at 29%. Such nodes may cross into the ACTIVE level with a small increase in the network load.

B_pER is another indicator of network performance which is calculated using the

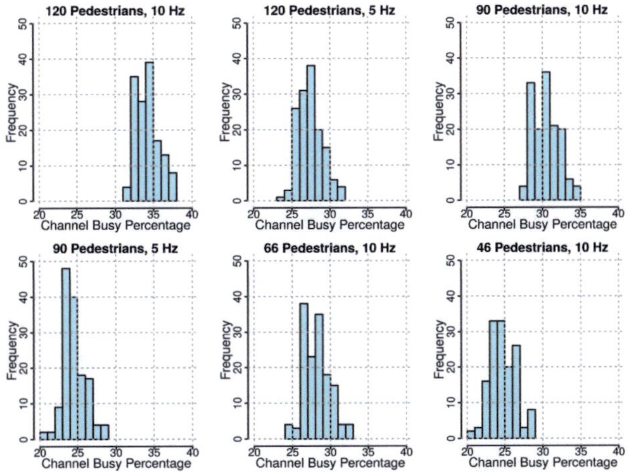

Figure 3.5: CBP Distribution [SKS17]

Equation (3.5) [SKS17]. We calculate the average of all B_pER values reported by vehicle nodes.

$$B_pER = \frac{num_{\text{dropped}}}{num_{\text{received}} + num_{\text{dropped}}} \tag{3.5}$$

where:

num_{dropped} = number of dropped beacons by each vehicle node
num_{received} = number of received beacons by each vehicle node

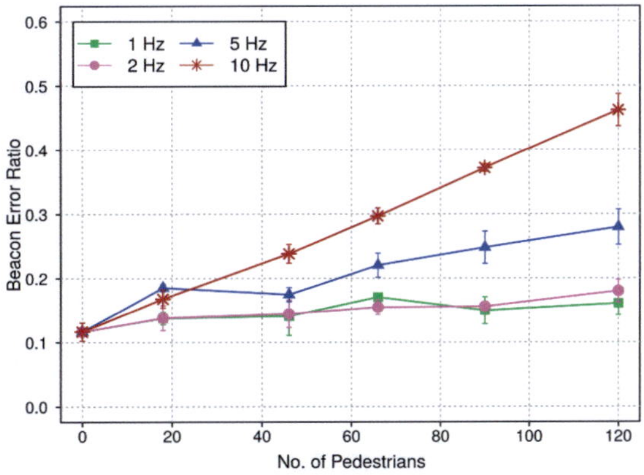

Figure 3.6: Beacon packet Error Ratio [SKS17]

Figure 3.6 shows the B_pER results. It shows that the B_pER increases with an increase in the number of pedestrian nodes as well as pedestrian beaconing periodicity. This trend in the results is expected. However, the increase in B_pER from 0.11 in the case of V2V with 0 pedestrians to 0.28 and 0.46 in the case of 120 pedestrians with 5 Hz and 10 Hz beaconing periodicity, respectively is large and underscores another challenge of reliability in V2X communication.

3.5.3.2 Quality of Service for Crucial V2P Communication

Our simulation scenario comprises two pairs of vehicles and pedestrians that are on the verge of collision as described in section 3.5.1. We measure the performance of the exchange of beacons between each pair in presence of the other V2V and V2P communication. This provides us insight into the QoS for crucial communication. We use the metric crucial Beacon delivery Ratio (B_cDR) which is given by equations (3.6) and (3.7) [SKS17].

$$B_cDR_{(\mathrm{V})} = \frac{num_{\mathrm{received(Vehicle)}}}{num_{\mathrm{sent(Pedestrian)}}} \tag{3.6}$$

where:

$num_{\mathrm{received(Vehicle)}}$ = number of beacons received by vehicle node
$num_{\mathrm{sent(Pedestrian)}}$ = number of beacons sent by pedestrian node

$$B_cDR_{(\mathrm{P})} = \frac{num_{\mathrm{received(Pedestrian)}}}{num_{\mathrm{sent(Vehicle)}}} \tag{3.7}$$

where:

$num_{\mathrm{received(Pedestrian)}}$ = number of beacons received by pedestrian node
$num_{\mathrm{sent(Vehicle)}}$ = number of beacons sent by vehicle node

Figure 3.7 shows the performance of crucial communication (along with standard deviation of values) for the pair in pre-crash scenario 1 (please refer to the figure 3.3). We can observe the trend of B_cDR decrease with the increase in the number of pedestrians and pedestrian beaconing periodicity. Also, the standard deviation is high for the pedestrian beaconing periodicity of 5 Hz and 10 Hz. It is also high for 2 Hz periodicity for 18, 66, and 90 pedestrians. However, as per Figure 3.4, CBP is at the RELAXED level for all the configurations with 2 Hz and 5 Hz periodicity. This indicates that crucial beacon delivery is unreliable even when the network load is low. B_cDR is very high in all cases with 1 Hz periodicity. However, beaconing with 1 Hz periodicity may not provide enough messages for various phases of crash prevention [SS19b].

Figure 3.8 shows the performance of crucial communication (along with the standard deviation values) for the pair in pre-crash scenario 2 (please refer to the

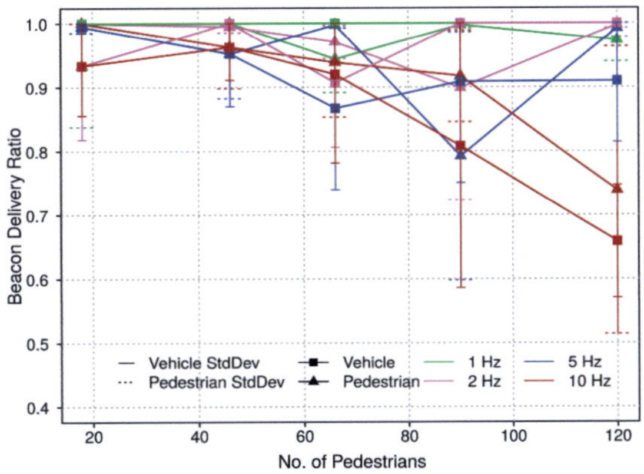

Figure 3.7: B_cDR for Pedestrian crossing in front of vehicle [SKS17]

figure 3.3). These results show the trends that are in line with the pre-cash scenario

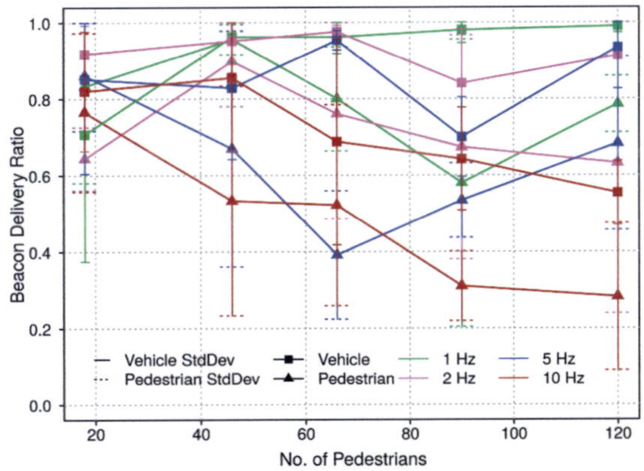

Figure 3.8: B_cDR for Pedestrian walking parallel to direction of vehicle [SKS17]

1. However, the standard deviation is even higher in this case showing worse reliability. Our further experiments with this case showed that the communication is reliable (100 % B_cDR) in the absence of other V2V and V2P communication. This pair may show worse communication performance due to the hidden node problem which is a common issue with CSMA/CA-based systems.

3.6 V2P Research and Development Tools

Researchers have leveraged the existing V2X tools to research V2P communication. In this section, we discuss the simulation tools that have been used for research and also, hardware tools that have been used for prototyping.

3.6.1 Simulation tools

3.6.1.1 OMNeT++, Veins, and SUMO

OMNeT++ is a modular and extensible communication network simulation library [omn]. It provides a platform for simulations using domain-specific model frameworks. Veins is such a V2X-specific model framework that works with OMNeT++ to provide V2X communication simulation capabilities [SGD11]. Veins supports DSRC/WAVE architecture, multi-channel functionality, and also, appropriate channel models. Simulation of Urban MObility (SUMO) is a mobility simulator that works with Veins in order to model various traffic scenarios [KEBB12].

3.6.1.2 Network Simulator-3

Network Simulator-3, or simply known as NS-3, is a discrete event simulator for network simulations [Nsn20]. It too supports WAVE architecture, multi-channel operations, and WSA functionality. It can take SUMO and Bonnmotion trace files as input for mobility modeling.

3.6.2 Hardware tools

3.6.2.1 V2X Hardware

There are few Original Equipment Manufacturers (OEMs) and developers of V2X solutions, such as Savari Networks, NeoGLS, who are currently supporting the development of V2P applications on smartphones. Ettifos Co. provides a prototype module specially designed to enable smartphones with V2X technology. However, there are currently no V2X-enabled smartphones or dedicated VRU devices available in the market.

3.6.2.2 Proof-of-Concept tools

The research community has developed the proof-of-concept V2P systems using various Commercial-Off-The-Shelf (COTS) smartphones in combination with

COTS OBUs. The smartphones allow various communication technologies to be used (cellular, Bluetooth, WiFi, etc.) and also to demonstrate the V2P concept. However, the maturity of such systems must be explored in the future when COTS smartphones are equipped with V2X.

3.7 Discussion

V2P communication is a special case of V2X communication as VRUs do not share many attributes with the vehicles. Due to this, the existing V2V solutions cannot be directly applied to the V2P systems. Also, our evaluation shows that presence of V2P communication with varying number of pedestrians and varying periodicity of beaconing have different effects on the V2V and V2P communication. The effects of enabling V2P on existing V2V communication have to be minimized. The periodicity of pedestrian beaconing is required to be optimized which ensures minimization of effect on V2V as well as enough availability for crash prevention phases. This means, just enabling V2P is not enough. The system needs to take a holistic approach, minimize the effect on V2V, and provide reliability for crucial V2P communication.

4 V2P System Design Framework

This chapter describes the proposed solution to the challenge of V2P system design. The proposed solution has been published in [SS19b]. This solution has also been provided as an input in the standardization efforts undertaken by the European Telecommunication Standards Institute under "Specialist Task Force 565 (ETSI STF 565)" which works on the specifications for the definition of C-ITS VRU service [SS19a, Eur19].

4.1 Introduction

As discussed in section 3.4.1, a V2P system has to be designed in line with the various aspects and configurations of the ITS systems. The V2P system design has to employ a systematic approach that would address these concerns. The V2P system design framework, proposed in this chapter, defines nine different categories for the design input. These categories are proposed after an extensive survey of the existing V2P system prototypes, traffic safety reports, and real-life pilot projects. Based on these categories, a case study of different configurations, with a focus on communication, was performed. This case study demonstrates the framework's utility in defining and understanding the design needs of the V2P system.

4.2 Design Framework

This section defines the nine different categories for the design input.

4.2.1 Types of VRUs

This category defines the target VRU group for which the V2P system is intended. The VRU characteristics have been discussed in section 3.3. We provide further details on VRU categories here.

- Pedestrians - This group can be further classified into following categories:

1. Adults - This category follows the typical pedestrian characteristics, such as walking speed, mobility patterns, etc. This category also includes distracted pedestrians, such as people listening to music, joggers, skateboarders, etc.
2. Children - This category may exhibit varying walking speeds and continuously changing trajectory patterns.
3. Seniors and physically disadvantaged - This subcategory may exhibit slow walking speeds and assisted walking (e.g. cane, wheelchair, guide animals, etc.)
4. Special Personnel - This involves people who are using the road for a special purpose, such as, road work, public safety, etc.

- Cyclists - This group can be further classified into the following categories:
  1. Human-powered cycles - These cyclists' typical traveling speed is 4.2 m/s (15 km/h).
  2. E-bikes - E-bikes may travel with a speed up to 25 km/h in Europe. In the USA, E-bikes may travel up to 32 km/h.
- Motorized Two Wheelers (MTWs) - This group is further divided into the following categories:
  1. mopeds - These MTWs travel up to 50 km/h. They are mainly found in urban and suburban areas.
  2. Motorcycles - These MTWs are the fastest category of VRUs. They are present in an urban area as well as on highways. In the urban area, they may travel with the typical urban speed of 50 km/h and on highways, they may travel up to 200 km/h.

Due to the varying characteristics of VRU groups, knowing the target VRU group and understanding their characteristics can help design the V2P system architecture better.

4.2.2 Pre-crash scenarios

Pre-crash scenarios depict the details of geographical characteristics and mobility patterns of the vehicle and VRU involved in the crash. The pre-crash scenarios vary by the different VRU groups. According to USA National Highway Traffic Safety Administration's (NHTSA) report on the V2P collision avoidance system, a large majority of pedestrian fatalities occur under only two types of pre-crash scenarios. Figure 4.1a shows a scenario where the pedestrian is crossing the road in front of the vehicle traveling in a straight line. This pre-crash scenario pertains to 88% of pedestrian fatalities that happened in the USA between 2005-2009 [CFM+13]. Figure 4.1b shows a scenario where the pedestrian is walking parallel to the direction of the vehicle (towards or away). 12% of estimated pedestrian fatalities pertain to this scenario. Also, according to another NHTSA report on V2P system analysis [SYN+16], the major part of the vehicle-pedestrian crashes happen on straight roads and away from the intersections. In the case of cyclists, the

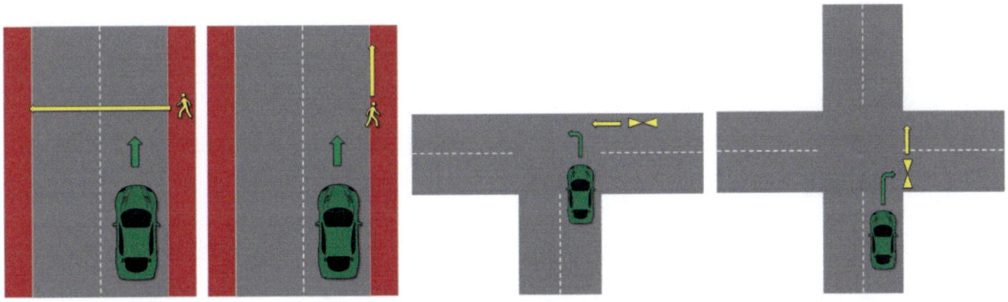

(a) Scenario 1 (b) Scenario 2 (c) Scenario 3 (d) Scenario 4

Figure 4.1: Various pre-crash scenarios: **(a)** Pedestrian crossing in front of the vehicle, **(b)** Pedestrian moving parallel to the vehicle, **(c)** Vehicle turning left into the cyclist's path, **(d)** Vehicle turning right into the cyclist's path [SS19b]

prominent pre-crash scenarios are different from those of pedestrians. Pre-crash scenarios for cyclists are shown in Figure 4.1c and 4.1d [MHM+13]. Figure 4.1c depicts a scenario where the vehicle is turning left into the cyclist's path. Figure 4.1d depicts a scenario where the vehicle is turning right into the cyclist's path where the cyclist is traveling straight while crossing the intersection. Also, according to the study performed in the VRUITS project, the major part of vehicle-cyclist crashes happen near a junction or intersection [MHM+13].

These varying pre-crash scenarios may pose different communication requirements on the V2P safety communication. These requirements may need to be addressed for an effective V2P system.

4.2.3 Mode of Communication

The mode of communication helps decide how the vehicles and VRUs should communicate with each other; either *directly*, *indirectly* or a combination of both (*hy-*

brid).

4.2.3.1 Direct mode

This mode involves direct communication of VRUs and vehicles i.e. without any intermediate infrastructure node. VRU devices and OBUs communicate directly with each other. This mode offers low latency and real-time delivery which are the stringent requirements imposed by safety-critical communication. However, to enable this mode, VRU devices and OBUs need to be equipped with the same communication technology (such as IEEE 802.11p) which poses challenges in deployment. This mode also requires VRU devices to be equipped with an *Information Processing Unit* (as shown in Figure 3.2) which may require accurate sensors and high computing power.

4.2.3.2 Indirect mode

This mode involves indirect communication of VRUs and vehicles i.e. through infrastructure. In this mode, VRU devices and OBUs may communicate through infrastructure over the same or different communication technologies. The role of the infrastructure may vary in this system. For example, the infrastructure may only be responsible to forward the messages or it may also process them and decide the intended recipient. It is necessary to understand the capabilities of the infrastructure, such as latency, scalability, before deciding to use it for V2P, especially for the dissemination of safety-critical information.

Multi-hop communication is a slightly different alternative of the *indirect* mode. In this case, intermediate nodes may be used to facilitate the communication. For example, an RSU at an intersection may re-broadcast the messages to improve the communication in an NLOS scenario. This type of communication may also be used to support heterogeneous V2P systems. For example, a VRU device may communicate with an intermediate node with one technology; this intermediate node then may re-broadcast this communication to other nodes over another technology.

4.2.3.3 Hybrid mode

Hybrid mode involves VRU and vehicle communication using *direct* as well as *indirect* mode. In this mode, VRU devices and OBUs may communicate using different modes under different scenarios. This overcomes the drawbacks of both modes in terms of deployment challenges as well as infrastructure capabilities.

However, a hybrid mode system may require the continuous assessment of the most suitable mode to use. It may also require well-defined functionality for both modes which increases the system complexity.

4.2.4 Type of Application

V2P applications can be broadly classified into two types. These types are discussed in this section.

4.2.4.1 Safety

Safety application focuses on improving VRU safety by designing various mechanisms to make vehicles aware of VRUs in real-time. The safety application may be designed for a specific VRU group, geographical location, or a specific type of vehicle. For example, V2P Collision Warning is a safety app for vehicles and all types of VRUs. Pedestrian in Signalized Crosswalk is a safety application designed to make public transport bus drivers aware of the pedestrians that are crossing at the intersection and are in the path of the vehicle [USD14].

4.2.4.2 Convenience

Convenience type of applications focuses on improving VRU travel efficiency by providing them various types of traffic-related services. The convenience applications too may be designed for a specific VRU group or geographical location. GLOSA for e-bikes, re-routing assistance, and traffic incident advisory are some examples of such services. These applications may also be categorized into Infrastructure-to-Pedestrian (I2P) applications instead of V2P.

The communication requirements of both types of applications are different. Safety applications require low latency, reliability, and real-time delivery in the communication. Convenience applications do not have such stringent requirements for communication.

4.2.5 Communication Technology

Communication technologies vary by their characteristics, such as communication range, latency, infrastructure requirement. These characteristics may have a direct effect on the V2P system performance and hence, it is important to choose the underlying communication technology carefully. This section discusses the potential of current communication technologies for V2P systems.

4.2.5.1 IEEE 802.11p

IEEE 802.11p technology has been specifically developed for V2X communication. As discussed in chapter 2, IEEE 802.11p has been used in DSRC and WAVE architectures. It supports V2X communication with low latency and up to 1 km of communication range. It does not require any infrastructure to operate although the deployment of RSUs is possible to support localized services. It can support communication reliably at high vehicle speeds. This technology can support the requirements of V2P safety applications. However, currently, 802.11p is not available commercially on any potential VRU device.

4.2.5.2 Cellular Technology

As discussed in section 3.4.1.2, cellular technology has been used to provide a proof-of-concept of V2P systems. However, current cellular technology generations, 4G and LTE, have a higher than required latency. It also requires the deployment of infrastructure which may not be cost-effective. Also, as the V2P applications inherently have a geographical scope, infrastructure nodes may require additional functionality to determine the geographical scope of message deliveries. This technology is commercially available in multiple devices which can serve as VRU devices and OBUs.

4.2.5.3 Wi-Fi

Although IEEE 802.11p is part of IEEE 802.11 Wi-Fi, it is discussed separately due to the inherent difference in operation as described in 2.3.1.1. Wi-Fi technology is ubiquitously available in various types of devices and hence, is easy to deploy. However, legacy Wi-Fi, such as, 802.11g, has a limited communication range (typically up to 100 meters) due to which the communication may not provide enough duration for crash prevention phases. Also, the legacy Wi-Fi technology requires that the devices must be attached to an Access Point (AP) to be able to communicate with each other. The attachment process may take up to few seconds which presents a big challenge due to highly mobile vehicles. This limitation of the legacy Wi-Fi can be overcome by the Wi-Fi Direct technology. However, the signal propagation properties of Wi-Fi Direct under vehicular mobility conditions remain largely unseen.

4.2.5.4 Localization using tags

Localization technology can be used in V2P systems to determine the distance between the vehicle and a VRU as discussed in section 3.4.1.2. This technology provides the distance estimation within the 100 m range. It uses a special transponder device as a VRU device which may be useful for the VRU groups that do not carry typical smartphones, such as, children. This technology may suffer from scalability challenges due to its support for a fixed number of transponders and TDMA communication.

4.2.5.5 Bluetooth

Bluetooth technology provides a platform for a localization service through iBeacon technology and for reliable data delivery through Peer-to-Peer (P2P) connectivity. Bluetooth provides a communication range of up to 50 m. Also, Bluetooth communication may not be suitable for highly dynamic vehicular movements. However, it may be suitable to provide connectivity for notifications. For example, it may provide connectivity between a smartphone and a Bluetooth-enabled headset to deliver an audio warning as a collision notification for VRU.

4.2.5.6 700 MHz ITS Band

700 MHz ITS band is used in Japan for V2X communication. This technology has been standardized by the Association of Radio Industries and Businesses (ARIB) in STD-T109 [ARI12]. The technology provides better channel access for infrastructure nodes which makes it a suitable candidate for infrastructure-assisted V2P communication (such as, heterogeneous or multi-hop communication as discussed in section 4.2.3). Also, due to its path loss characteristics, the technology provides better performance in urban environments [HKDS18].

4.2.5.7 802.15.4

This technology can be used in V2P systems for the detection of VRUs by vehicles. It can provide a communication range up to 80 m in NLOS scenario [LBKW13]. The VRU device can be in the form of a tag and is suitable for VRUs not carrying smartphones. This is suitable for systems where only the vehicle drivers need to be notified (no notification for the VRUs.)

4.2.6 VRU Devices

Varying characteristics of VRUs imply that the usability is not the same for all groups of VRUs and so there is no one-size-fits-all solution when it comes to VRU devices. For example, smartphones as VRU devices are suitable for typical adult pedestrians but may not be suitable for children or physically disadvantaged people. This section discusses the potential of various VRU devices.

4.2.6.1 Smartphone

Current COTS smartphones are already equipped with various sensors, such as IMU, GPS, and communication technologies, such as cellular, Wi-Fi, Bluetooth, etc. They also are equipped with powerful processors that may provide enough computing capabilities for processing the V2P messages in real-time. Apart from this, smartphones are equipped to provide audio-visual and haptic notifications. This may make smartphones as one of the most widely acceptable VRU devices for pedestrians. However, smartphone's usability to provide audio-visual warnings when they are inaccessible (such as, in the pocket of pedestrians or a cyclist) needs to be researched further. Also, children and seniors may not use smartphones. MTWs may have a dedicated VRU OBU and may not use the smartphone.

4.2.6.2 Helmet

Cyclists and MTWs may use helmets as a VRU device. Helmets need to be equipped with the necessary components to be a suitable VRU device. Helmets may also work in tandem with smartphones and provide functionality for warning delivery mechanisms.

4.2.6.3 Specialized tag

A V2P system may also use specialized tags to make vehicles aware of the VRU's presence. A vehicle OBU may detect the presence of the tag and also, estimate the distance to the tag. Tags are especially useful for children, seniors, and physically disadvantaged people as they can be attached to their accessories (backpacks, wheelchairs, etc.).

4.2.7 Role of VRU Devices

This input defines the participation functionality of the VRU device in V2P communication. This participation can be defined in the following ways:

4.2.7.1 Active

This role involves the VRU device communicating *actively* with the vehicles i.e. the VRU device broadcasts the safety messages periodically to its surroundings. In this role, the VRU device needs to be equipped with positioning technology (GPS, Galileo, etc), communication technologies (DSRC, Cellular, etc.), and also, IMU sensors. With powerful devices, such as smartphones, the VRU device may perform an active role in V2P communication. This role is useful when it is recommended for vehicles to be aware of the VRU presence in advance, such as, in NLOS scenarios.

4.2.7.2 Passive

Under this role, the VRU devices do not send safety messages periodically. The device may only 'listen' to the safety messages from vehicles and alert only the VRU when the crash is imminent. Another way of passive participation is that the VRU device may transmit a 'reply' message only when it detects the presence of a vehicle (through the received safety message) or when it predicts a crash with the vehicle. In this mode, the vehicle becomes aware of the VRU's presence only if the VRU device sends its information. Also, the instance of awareness may be delayed when compared to that of the *active* mode. This mode is also useful for I2P convenience applications, such as receiving traffic advisory, and traffic light phase information.

4.2.8 Notification Recipients

As discussed in section 2.4.2.2, there are multiple V2P safety applications that notify about the collision, crossing paths, etc. The safety applications are designed for different VRU groups using different VRU devices which may vary in their capabilities. For example, a smartphone and an MTW OBU have different capabilities. These factors require that the V2P system must design warning mechanisms with the appropriate end-users in mind.

- *Driver-only*: In this mechanism, when a vehicle-VRU collision is predicted to happen, only the driver of the vehicle is notified. This mechanism is useful when the target VRU group is either children, seniors, or physically disadvantaged people. Examples of notifications to the driver are in-vehicle signage or haptic warning.

- *VRU-only*: In this mechanism, when a vehicle-VRU collision is predicted to happen, only the VRU is notified. Examples of notifications to the VRU are in-dash signage for MTW, audio-visual and haptic warning through the smartphone, etc.

- *Driver* and *VRU*: In this mechanism, when a vehicle-VRU collision is predicted to happen, both VRU and driver are notified.

In the case of V2P convenience applications, only VRUs may get notified as these applications focus on providing information to VRUs.

4.2.9 Response Time Requirement

This input is for the V2P safety application which defines the requirement of the available response time. Available Response Time (ART) is defined as the duration between the estimated crash time and the times of the first contact between the vehicle and VRU. Due to varying pre-crash scenarios and VRU characteristics, the available response time may differ significantly, and hence, it is necessary to understand the ART for each VRU group.

4.3 Evaluation of Design Framework

The design inputs of the framework also describe the design choices of various components of the V2P system. It is necessary to understand the effects of these design choices on the performance of the V2P system. This section discusses a case study of different pre-crash scenarios with different design inputs.

4.3.1 Concept

As discussed in 3.2.2, the V2P safety system works in 3 different phases. *Detection* phase or the first contact must occur sufficiently early in order to provide enough duration for the other two phases, (*tracking and prediction* and *action*). Enough duration for these two phases also implies that a sufficient number of safety messages are exchanged. This can be studied by employing different *roles of VRU devices* i.e. by using *active* and *passive* modes. For evaluation purpose, these two modes are defined as follows:

Active Mode

When the VRU device assumes the *active* role in V2P communication, it may generate safety messages periodically. As discussed in section 3.4.2, this periodicity may be determined based on the VRU context. For this evaluation study, it is set to 2 Hz. This is based on the assumption that as the VRUs travel slowly compared to the vehicle, this periodicity may be able to provide the updates from VRUs that are sufficient for the crash prediction phases.

Passive Mode

When the VRU device assumes the *passive* role in V2P communication, it may transmit the safety message only after the detection of the vehicle's message or when it detects that the crash is imminent. For evaluation, the latter approach is adopted where the VRU device waits for two seconds after it receives the first message from the vehicle and determines that the crash is imminent. It then starts transmitting the safety messages every two seconds.

Another measure for the duration, after the first contact, is the *Available Response Time* which is proposed as part of the input *Response Time Requirements.* ART provides insights into the response time for different pre-crash scenarios.

The following list describes the design choices that have been chosen to understand their effect on the performance of the V2P system:

- Type of VRUs: Pedestrians and cyclists are used for evaluation as they differ in their mobility and pre-crash scenarios.
- Pre-crash scenarios: 4 prominent pre-crash scenarios have been chosen for evaluation which incorporates pedestrians and cyclists.
- Type of application: The purpose of the evaluation is to assess the performance of V2P safety communication.
- Communication technology: The WAVE architecture is used for exchanging safety messages which allows evaluating IEEE 802.11p based V2P communication. As the IEEE 802.11p technology has been widely researched for V2V and V2I communication, this evaluation provides insights into the technology's application to V2P.
- Mode of communication: The choice of IEEE 802.11p allows V2X devices to participate in communication in *direct* mode.

- Role of VRU device: The evaluation compares the *active* and *passive* roles of VRU devices.
- Response time requirement: The evaluation measures the ART for various scenarios.

4.3.2 Scenario

For evaluation purposes, a T-junction from an urban area is considered. This T-junction sees the traffic of vehicles, pedestrians, and cyclists. It incorporates the 4 pre-crash scenarios that are depicted in 4.1. Figure 4.2 shows the T-junction with 4 pairs of vehicles and VRUs that are on the verge of collision. The two roads forming the T-junction contain two lanes each in both directions. The traffic at the intersection is controlled by the traffic lights. The vehicles are inserted at the far end of the road. The roads have footpaths on both sides which are used by pedestrians and cyclists. The vehicles and VRUs use DSRC communication technology to transmit safety messages.

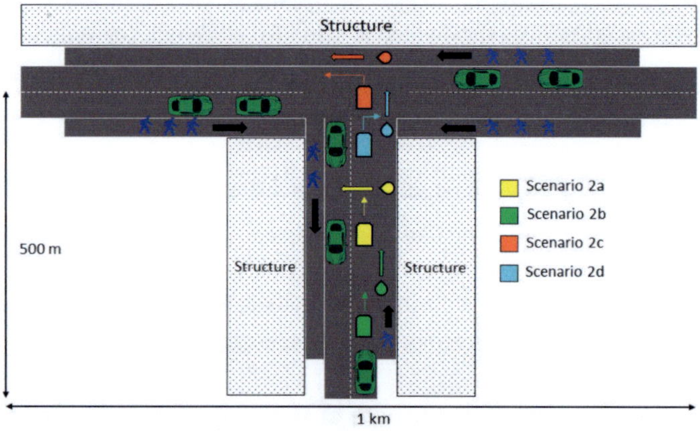

Figure 4.2: Evaluation scenario [SS19b]

4.3.3 Simulation Environment

The evaluation is performed using OMNeT++, Veins, and SUMO. Vehicles and pedestrians are inserted in SUMO every 2 s and 1.6 s, respectively. The transmission power of all the nodes is set to 20 mW which allows a communication range up to 400 m. The signal propagation is simulated using the TwoRayInterferenceModel path loss model. Table 4.1 lists the simulation parameters.

Parameter	Value
Road length	1 km x 500 m
No. of vehicles	120–150
Max. vehicle speed	13.89 m/s = 50 km/h
No. of pedestrians	102
Max. pedestrians speed	1.5 m/s
No. of bicycles	1
Max. bicycle speed	4.3 m/s
Vehicle transmission power	20 mW
VRU transmission power	20 mW
Data rate	6 Mb/s
Vehicles beacon periodicity	10 Hz
VRU beacon periodicity (for active mechanism)	2 Hz
Beacon length	1024 bits

Table 4.1: Simulation parameters [SS19b].

Table 4.2 shows the simulation warm-up period, crash time, and simulation run time for each pre-crash scenario. The warm-up period indicates the time required for the simulation to bring the T-junction into a typical traffic situation. Crash time for the given pre-crash scenario is defined as the simulation time when the vehicle and VRU collide with each other.

Scenario	**Warm-up** (in s)	**Crash time** (in s)	**Simulation length** (in s)
2.a	30	33.5	33.5
2.b	11	44	44
2.c	45	49	49
2.d	10	48	48

Table 4.2: Scenario-specific parameters [SS19b]

Each pre-crash scenario configuration is run three times with different seeds and the mean values of gathered data are used for the evaluation.

4.3.4 Evaluation

The measure of duration for *tracking and prediction* and *action* phases is given using *Available Response Time*. It is calculated using following formula:

$$ART = CT - FBT \tag{4.1}$$

where:

ART = Available Response Time
CT = Crash Time
FBT = First Beacon Time

For a pair of vehicle and VRU that are on the verge of collision, FBT denotes the timestamp of the *detection* of the first message (beacon) received by the vehicle from the VRU. CT denotes the timestamp when the pair would crash into each other.

Another measure affecting the *tracking and prediction* phase is the number of safety messages received by the vehicle from the corresponding VRU. This is affected by the periodicity of the VRU beacons as well as the contact duration. Table 4.3 shows ART and number of received messages values measured for 4 pre-crash scenarios with *active* and *passive* VRU device roles.

Scenario	**Available Response Time** (before crash, in s)	**No. of Received Messages** (from VRUs)
2.a		
Active	2.13	5
Passive	0.39	1
2.b		
Active	31.7	62
Passive	29.96	13.33
2.c		
Active	3.65	7.33
Passive	1.27	1
2.d		
Active	37.65	74
Passive	35.93	16.67

Table 4.3: Results [SS19b]

The results reveal interesting insights for various design inputs. The ART values varies by the pre-crash scenario and it is lowest for the scenario 2.a for both, *active* and *passive* roles. Scenarios 2.a and 2.c have the ART values that are below the 8 seconds crash warning threshold as specified by SAE J2945/9. Scenarios 2.b and 2.d have sufficiently high ART values which would suffice the 8 seconds requirement. Another characteristic of ART is that it is higher in *active* role than in *passive* role in all scenarios. This provides more time for the *tracking and prediction* and *action* phases in *active* role. Further evaluation of scenario 2.a shows that the low ART value is due to the short distance (< 4 m) that the pedestrian travel before it crashes into the vehicle. Evaluation of scenario 2.c shows that the low ART value is due to the NLOS between the cyclist and the vehicle. The signal between the vehicle and the cyclist is blocked by the obstacle (building).

Similar to ART, number of beacons are more in *active* role than in *passive* role. The results also reveal that under scenarios 2.a and 2.c, the vehicle receives only 1 safety message from the corresponding VRUs. This message would be sufficient only for the *detection* phase and there are no messages available for the other two phases. The vehicles receive a sufficiently large number of messages under scenarios 2.b and 2.d. The number of beacons transmitted by VRUs is also affected by the periodicity of beacon transmission. This periodicity may be increased in order to have more beacons transmitted.

4.4 Discussion

The V2P system design framework proposes a holistic and systematic approach towards the development of the V2P system. It categorizes the design inputs into nine classes and discusses the possible design choices for each of these classes. As the evaluation shows, the design inputs are correlated to each other and affect the overall V2P system performance.

5 Network Congestion Control in V2P Networks

This chapter describes the proposed solution to the V2X network congestion issue caused by VRU-generated safety messages. It describes the system model of the proposed solution, details of evaluation of the model, and results of the evaluation. The proposed solution has been submitted for publication [SS20].

5.1 Introduction

When VRUs are present in large numbers in a given area, the VRU-generated safety messages can easily congest the V2X network. As discussed in section 3.4.2, a mechanism that can reduce the impact of VRU-generated safety messages on overall V2X communication is needed.

This chapter introduces a network congestion mitigation scheme called Multi-channel Clustering-based Congestion Control for Vehicle-to-Pedestrian Communication or MC-COCO4V2P in short. This scheme focuses on network congestion caused by a large number of pedestrians in a given area. It employs a clustering mechanism to group the pedestrians together for V2P communication. This reduces the number of pedestrians actively transmitting the safety messages which in turn reduces the load on the V2X network. The mechanism also employs a transmission power control mechanism which uses different transmission power levels for intra-cluster and safety message communication. This improves the power consumption efficiency of the pedestrian devices. The MC-COCO4V2P approach is based on the DSRC/WAVE architecture. However, it can be implemented using any technology that allows ad-hoc direct communication and also, allows separation of control and safety messages.

5.2 Multi-channel Clustering-based Congestion Control

5.2.1 Concept

MC-COCO4V2P mechanism consists of two major components: a) Clustering algorithm, and b) Transmission Power Control (TPC) mechanism. The clustering algorithm helps mitigate network congestion while TPC helps improve the energy efficiency of the pedestrian device.

Pedestrian devices, such as smartphones, are equipped with positioning systems (GPS/GLONASS/Galileo) and IMU (accelerometer/gyroscope). These sensors provide the location, speed, and direction information about the pedestrian. MC-COCO4V2P employs a *proactive* clustering mechanism which uses distance, direction, and battery life information, extracted from the pedestrian device, to form clusters of pedestrians. Pedestrian devices exchange this information among themselves to choose the role of either Cluster Head (CH) or Cluster Member (CM). CH is responsible to perform cluster maintenance as well as maintain V2P communication on behalf of the cluster. MC-COCO4V2P uses two separate MAC channels for V2P safety messages and Pedestrian-to-Pedestrian cluster control messages. This ensures that the control messages do not increase the load on the channel used for safety communication. Figure 5.1 illustrates an example of clustering of pedestrians in a V2X network.

Further, MC-COCO4V2P's transmission power control mechanism is employed to improve the power consumption for the cluster operations as well as the V2P safety communication. Both of these communication types vary in their requirements of communication range. Pedestrian-generated safety messages have to reach over a few hundred meters to achieve a safe braking distance (for vehicles). For MC-COCO4V2P's cluster communication, the range can be limited to only up to a few meters. Due to such difference, MC-COCO4V2P's TPC mechanism differentiates between the transmission power required for the cluster and the safety communication. This provides the following advantages:

1. Reduction of the inter-channel interference,
2. Decrease in the channel load for intra-cluster communication,
3. Improvement in the power consumption of the pedestrians' devices.

TPC mechanism uses two different transmission power levels: low transmission power to achieve short delivery distances for intra-cluster communication and high transmission power to achieve longer delivery distances for safety messages.

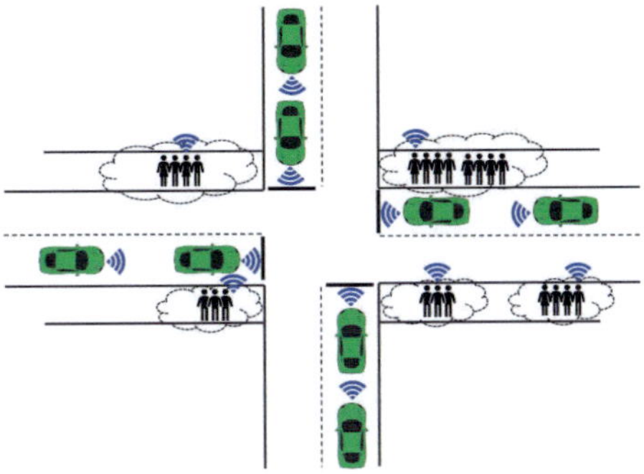

Figure 5.1: Clustering of pedestrians [SS20]

5.2.2 Assumptions

MC-COCO4V2P approach makes a few assumptions which are listed below:

- All pedestrians carry a DSRC-capable VRU device such as smartphone,
- All pedestrians participate in V2P communication,
- The VRU devices are equipped with single-radio DSRC transceivers,
- The VRU devices are synchronized using GPS,
- The VRU devices are capable of multi-channel operation as shown in figure 2.5 and alternate between the CCH and SCH 174,
- The VRU devices use CCH and SCH 174 for safety messages and cluster control messages, respectively (please refer to section 2.3.1.2 for WAVE channel details).
- Vehicles are tuned to CCH for exchange of safety messages.

5.2.3 Clustering Architecture

MC-COCO4V2P clustering architecture relies on 4 different components which are discussed in this section.

Cluster Notations	Description
Cluster Head	CH
Cluster Member	CM
CLUSTER_INVITE	Cluster Awareness Announcement
CLUSTER_JOIN	Cluster Join Request
CLUSTER_ACK	Cluster Acknowledgment
CLUSTER_UPDATE	Cluster Update Request
CLUSTER_DISSOLVE	Cluster Dissolve Operation
neighborScanningTimer	Timer for Neighbor Discovery
CSM	Cluster-Aware Safety Message
ClusterRadius	Maximum distance between nodes for cluster formation
BATTERY_DIFF_THRESHOLD	Battery Difference Threshold parameter

Table 5.1: Cluster Notations

5.2.3.1 Clustering Algorithm

As discussed in section 2.4.2, the clustering process consists of various operations. Similarly, the MC-COCO4V2P clustering algorithm too carries out various operations. The cluster-specific notations are summarized in Table 5.1.

1. Neighbor Discovery

This is an initial phase of clustering for a given pedestrian device where the pedestrian is not part of any cluster. In this phase, the pedestrian device listens for cluster announcements (CLUSTER_INVITE) on SCH till the expiry of a timer *neighborScanningTimer*. It stores the received CLUSTER_INVITEs for the selection of an appropriate CH till the expiry of *neighborScanningTimer*. It also transmits periodic safety messages on CCH.

2. Cluster Head Selection

This phase starts after the expiry of *neighborScanningTimer*. The pedestrian device scans through all of the received CLUSTER_INVITEs during the *Neighbor Discovery* phase to find a suitable CH based on the CH selection criteria. If a suitable CH is found, the clustering process proceeds to step 3. If a suitable CH is not found, then the device assumes the CH role itself and proceeds to step 4.

3. Cluster Affiliation

After the selection of a suitable CH, the pedestrian device sends a request to this

CH for joining the cluster using the CLUSTER_JOIN message. Upon reception of CLUSTER_JOIN, the CH adds the information of the requesting device to its cluster information table and sends an acknowledgment (CLUSTER_ACK) for successful addition. Upon reception of this CLUSTER_ACK, the requesting device assumes the CM role and proceeds to step 4.

4. Cluster Maintenance

Cluster maintenance operation involves different roles for CM and CH which are described below.

- As a CM: CM node sends periodic updates (CLUSTER_UPDATE) of its location and direction to the CH of the affiliated cluster.
- As a CH: CH broadcasts the cluster announcements (CLUSTER_INVITE) periodically to indicate the presence of the cluster. It also monitors the CLUSTER_UPDATE messages from the member nodes and updates the neighbor table accordingly. If the CH finds a node that is more suitable to be CH then it may also dissolve the cluster.

5.2.3.2 Cluster Head Selection Criteria

MC-COCO4V2P clustering algorithm selects the best CH based on the direction, distance, and battery charge. A node is chosen to be a CH by other nodes only if all of the following criteria are met:

- Potential CH's direction of travel is the same as that of the node
- Distance between the potential CH and the node is $\leq$ ClusterRadius
- Potential CH's battery charge is higher than that of the potential CM by at least BATTERY_DIFF_THRESHOLD.

BATTERY_DIFF_THRESHOLD is a threshold battery charge difference level. This makes sure that there no ping-pong effect in CH selection (no frequent CH changes). If a node cannot find any CH that meets all of the criteria above then the node itself assumes the CH role.

5.2.3.3 Cluster-aware Safety Message Delivery

In MC-COCO4V2P, the CH is responsible to maintain V2P communication on behalf of the cluster. This requires the CH to send safety messages that would make vehicles aware of the cluster. This is achieved through a newly defined safety message called Cluster-aware Safety Message (CSM). CSM encompasses the information about the cluster, such as speed, location, and number of CMs in that cluster. The *Vehicle Size* field is used as part of the CSM which is modified to mark the boundaries of the cluster. This also maintains the backward compatibility for the vehicles that cannot decode the CSMs. The CH always broadcasts the CSMs with 10 Hz periodicity. CSMs are broadcast on CCH.

5.2.3.4 Cluster Messages

MC-COCO4V2P defines various message formats to facilitate the clustering operations and V2P communication. These messages are described below.

- CLUSTER_JOIN - Control message sent by a node to potential CH requesting to join the cluster.

$$< NodeID,\ ClusterHeadID,\ NodeLocation,\\ NodeDirection,\ BatteryCharge >$$

- CLUSTER_INVITE - Control message transmitted by CH to announce the presence of a cluster.

$$< ClusterHeadID,\ ClusterHeadLocation,\\ ClusterDirection,\ BatteryCharge >$$

- Cluster-aware Safety Message (CSM) - Safety message sent by CH to the vehicles

$$< ClusterPosition,\ ClusterBoundary,\\ ClusterDirection,\ ClusterMemberCount >$$

5.2.4 Transmission Power Control

MC-COCO4V2P's TPC mechanism aims to improve the power consumption of pedestrian devices for V2P safety and cluster communication. V2P communication involves pedestrian devices actively exchanging safety messages with the vehicles. As the pedestrian devices need to receive the messages from the surrounding vehicles, their receivers need to be always on. This limits the scope of the improvement in power consumption largely to the transmitter side. For this reason, MC-COCO4V2P too focuses on power savings at the transmitter side. The proposed TPC mechanism uses a *cost function* [GCNB01] to analyze and develop the model of *Power Cost Function.*

The *Power Cost Function* is based on the power consumed by the pedestrian device for each message (safety and control), the frequency of the messages, and the time duration of the communication. Let $\mathrm{P_{HS}}$ and $\mathrm{P_{LC}}$ be the power consumed by the pedestrian device for the transmission of one safety message and one cluster message, respectively. As the communication range required for cluster message is less than that of safety message, $\mathrm{P_{LC}} < \mathrm{P_{HS}}$.

5.2.4.1 Power consumption without clustering

When pedestrians are not clustered for optimization purpose, the pedestrian device may transmit only safety safety messages. The total power consumed, $\mathrm{P_T}$, for such safety message transmission can be give by equation 5.1.

$$P_{\mathrm{T}} = P_{\mathrm{HS}} \cdot F_{\mathrm{S}} \cdot T \tag{5.1}$$

where:

F_{S} = periodicity of safety messages
T = time duration of safety communication

5.2.4.2 Power consumption with clustering

When the pedestrians are clustered together, the pedestrian device transmits safety messages during the formation of a cluster (Neighbor Discovery phase). It also exchanges cluster control messages during various stages of the cluster. Also, the pedestrian device may assume either a CM or a CH role in the clustering process which may also impact its power consumption.

Among all the clustering operations, the *Cluster Affiliation* operation involves only a one-time exchange of CLUSTER_JOIN/CLUSTER_ACK messages and hence, does not have a significant impact on the power consumption. For the other operations, the *Power Cost Function* can be given as below.

- Neighbor Discovery - In this phase, the pedestrian device may transmit the safety messages for the duration of *neighborScanningTimer* while it waits for the CLUSTER_INVITE messages. The power consumed during this phase,

P_{ND}, can be given by equation 5.2.

$$P_{ND} = P_{HS} \cdot F_S \cdot neighborScanningTimer \tag{5.2}$$

where:

F_S = periodicity of safety messages
$neighborScanningTimer$ = timer for Neighbor Discovery phase

- Cluster Maintenance - Pedestrian devices may assume either a CM or a CH role during the cluster operation. As a CM, the pedestrian device transmits CLUSTER_UPDATE periodically to maintain its membership of the cluster. Power consumption in the CM role, P_{TCMA_CM}, over the time T s can be determined by the equation 5.3.

$$P_{TCMA\_CM} = P_{LC} \cdot F_{CU} \cdot T \tag{5.3}$$

where:

F_{CU} = periodicity of CLUSTER_UPDATE messages

As a CH, the pedestrian device transmits CLUSTER_INVITE periodically to announce the presence of the cluster. Power consumption in the CH role, P_{TCMA_CH}, over the time T s can be determined by the equation 5.4.

$$P_{TCMA\_CH} = P_{LC} \cdot F_{CI} \cdot T \tag{5.4}$$

where:

F_{CI} = periodicity of CLUSTER_INVITE messages

- Cluster-aware Safety Message Delivery - The pedestrian device with CH role transmits CSMs on behalf of the cluster. As CSMs are safety messages, they are transmitted with the power P_{HS}. Power consumed for this operation, P_{TCS}, over time T s, can be determined by the equation 5.5.

$$P_{TCS} = P_{HS} \cdot F_S \cdot T \tag{5.5}$$

where:

F_S = periodicity of CSM safety messages

The aggregate power consumed by the pedestrian device with a CM role, P_{TCM}, can be given by equation 5.6

$$P_{TCM} = P_{ND} + P_{TCMA\_CM} \tag{5.6}$$

The aggregate power consumed by the pedestrian device with a CH role, P_{TCH}, can be given by the equation 5.7

$$P_{TCM} = P_{ND} + P_{TCMA\_CH} + P_{TCS} \tag{5.7}$$

The improvement in the power consumption may be observed by fulfilling the criteria: a) $P_{TCM} < P_T$, and b) $P_{TCH} \lesssim P_T$.

5.3 Performance Evaluation

This section discusses the evaluation of MC-COCO4V2P to verify the mitigation of the network load caused by pedestrian-generated safety messages and the improvement in the power consumption of pedestrian devices.

5.3.1 Scenario

MC-COCO4V2P is implemented and evaluated using IEEE 802.11p-based DSRC-WAVE protocol stack. DSRC/WAVE protocol stack allows the multi-channel operation which is required to separate the safety and control messages (cluster message). It also enables the network operation in an ad-hoc mode which allows the pedestrian nodes to exchange control messages without network association (unlike Wi-Fi). TPC mechanism leverages the capability of a WAVE-based V2X network to set the transmission power on a per-messages basis.

A typical urban intersection scenario is considered for evaluation. The scenario sees the vehicles and pedestrian traffic. Figure 5.2 illustrates the scenario used for the evaluation. The intersecting road segments are 500 m long each and contain two lanes in each direction. The vehicles enter the simulation at one end of each segment and travel straight to the other end where they exit the simulation. The scenario contains eight footpath segments that are used by pedestrians to travel either towards or away from the intersection as depicted. The scenario also contains the building blocks alongside the road segments which help simulate the urban environment affecting the radio propagation.

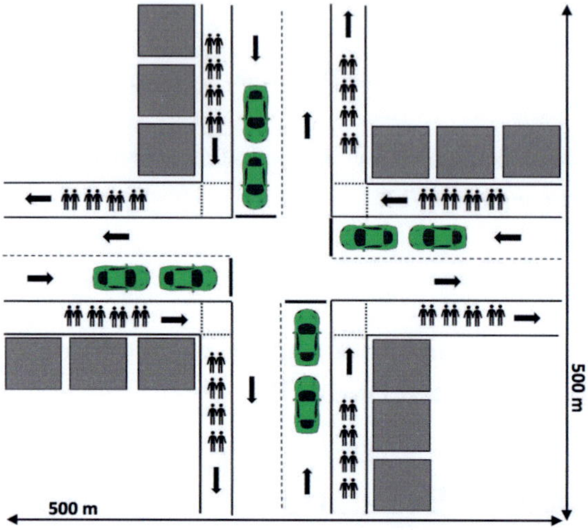

Figure 5.2: Evaluation scenario [SS20]

The vehicles and the pedestrians travel at the typical urban speed of 13.89 m/s (50 km/h) and 1.5 m/s, respectively. The vehicles and pedestrians exchange the safety messages on CCH. The pedestrian nodes use SCH 174 for the exchange of cluster information. The vehicle nodes are tuned to CCH in the continuous access mode of operation. The pedestrian nodes operate in the alternating access mode. They alternate the channel access between the CCH and SCH. Different channel access modes already have been depicted in figure 2.5. The vehicles broadcast the safety messages with 10 Hz periodicity. The pedestrians' safety message periodicity is selected from 2, 5, and 10 Hz. The CSM periodicity is set to 10 Hz which means CHs always transmit the CSMs with 10 Hz periodicity. This improves the reliability of safety messages transmitted on behalf of the cluster. CHs and CMs transmit the cluster maintenance messages (CLUSTER_INVITE and CLUSTER_UPDATE) with 2 Hz periodicity. The cluster radius is set to 5 m which enables the formation of clusters of pedestrians that are walking close to each other. This also allows the TPC mechanism to use low transmit power for intra-cluster communication. Table 5.2 shows the cluster parameter values used in the simulation.

Cluster Parameters	Value
Cluster Radius	5 m
CLUSTER_INVITE Periodicity	2 Hz
CLUSTER_Update Periodicity	2 Hz
CSM Periodicity	10 Hz
neighborScanningTimer	1.5 s
BATTERY_DIFF_THRESHOLD	3%

Table 5.2: Cluster Parameters [SS20]

5.3.2 Simulations

MC-COCO4V2P implementation and the scenario design are achieved using the OMNeT++, Veins, and SUMO framework. The details of this framework are described in section 3.6.

The number of vehicles in the scenario is set to 140. The number of pedestrians is varied from 80 to 240. This allows us to study the performance under a varying number of pedestrians and hence, varying network load and pedestrian density.

TPC mechanism implementation uses two Tx power levels for pedestrian-generated safety messages (20 mW) and cluster messages (1 mW). As the simulation framework does not offer a mechanism to measure the power consumption on a per-message basis, this work implemented a power cost calculator that allows measuring the desired power consumption of pedestrian devices. For the simulation purpose, "one *battery unit* is defined as an amount of power used when a message was transmitted using 1 mW of Tx power" [SS20]. The power cost calculator relies on consumed *battery units* to calculate the power cost function of the TPC mechanism. The power cost calculator of each pedestrian device is initialized with *InitialBatteryUnits* which is an initial battery charge. The *InitialBatteryUnits* is generated randomly during the initialization of the pedestrian node (when the node enters into the simulation). The power cost calculator monitors the Tx power of each outgoing message, calculates the consumed *battery units* and then determines the remaining battery charge. This method helps determine the total number of *battery units* consumed during the evaluation.

The simulation employs *SimpleObstacleShadowing* and *TwoRayInterference-Model* modules as path propagation models to estimate the signal propagation in urban environment. The simulation warm-up period is 38 s which allows the scenario to get into a typical intersection traffic state. The simulation length is 58 s allowing the measurement window to be 20 s. Each simulation configuration is run independently three times. The data gathered over three runs is averaged and then further processed for result analysis. Table 5.3 provides details of simulation parameters.

Simulation Parameters	Value
Road length	500 m
Road layout	Two-way, two-lanes
No. of vehicles	140
Max vehicle speed	13.89 m/s = 50 km/h
No. of Pedestrians	80, 130, 180, 240
Max pedestrian speed	1.5 m/s
Vehicle beacon periodicity	10 Hz
Pedestrian beacon periodicity	2,5,10 Hz
Vehicle Tx Power	20 mW
BSM size	135 bytes
Cluster safety message size	147 bytes
Cluster management message size	133 bytes
Pedestrian Tx power (safety message)	20 mW
Pedestrian Tx power (intra-Cluster message)	1 mW
Pedestrian Tx power (CSM)	20 mW

Table 5.3: Simulation parameters [SS20]

5.3.3 Evaluation

As MC-COCO4V2P aims to reduce the network congestion and power consumption of the pedestrian devices, its performance can be evaluated by its effect on these elements. Consequently, the following metrics are used for performance evaluation of the clustering algorithm and TPC mechanism.

1. Channel Busy Percentage

Channel Busy Percentage (CBP), also called Channel Busy Ratio (CBR), is the indicator of channel load in the V2X network. As the clustering algorithm's primary purpose is to reduce the network load caused by pedestrian-generated messages, CBP is measured only for the vehicular nodes. The CBP of vehicular nodes, when the pedestrians are not clustered, is compared against that of when the pedestrians are clustered. This provides insights into the impact of the clustering algorithm on the network load. CBP is calculated using the equation 5.8.

$$CBP = \frac{t_{\text{ChBusy}}}{t_{\text{CBP}}} \times 100\% \tag{5.8}$$

where:

t_{ChBusy} = duration of the time when the channel was busy
t_{CBP} = vehicle's lifetime during the measurement window

The vehicle's lifetime is the duration of time when the vehicle is present in the simulation. The vehicle may be present only for a fraction of its lifetime during the measurement window. Only this fraction of its lifetime is considered for measurement. CBP is calculated for all the vehicles individually and then the mean value is considered for analysis.

2. Number of pedestrian-generated beacons

MC-COCO4V2P's clustering algorithm groups the pedestrians together so that only the CHs broadcasts the CSMs (safety messages). This directly affects the total number of pedestrian-generated beacons. When the pedestrians are not clustered, the total number of beacons is the individual safety message (BSMs) generated by pedestrians. When the pedestrians are clustered together, the total number of beacons is the total of the CSMs sent by CHs and the BSMs sent by individual pedestrians (that were sent during Neighbor Discovery or when the pedestrian is not part of any cluster). The total number is calculated by adding the numbers reported by each pedestrian node.

3. Power consumption of pedestrian devices

The TPC mechanism uses two different power levels for safety communication and intra-cluster communication. This directly affects the power consumed by pedestrian devices. The power consumption is measured using the *battery units* by each pedestrian device and then the mean value is considered for analysis. Apart from the mean power consumption, the maximum power consumption reported by pedestrian devices is also considered.

Figure 5.3 shows the comparison of the CBP reported by vehicle nodes for clustered and un-clustered configurations. CBP increases with the number of pedestrians and the periodicity of pedestrian-generated safety messages in the un-clustered configuration. This is expected as this work has shown before in figure 3.4. However, for all clustered configurations, CBP values remain relatively at the same levels - between 9%-10%. This is due to the reduced number of pedestrians transmitting the safety messages.

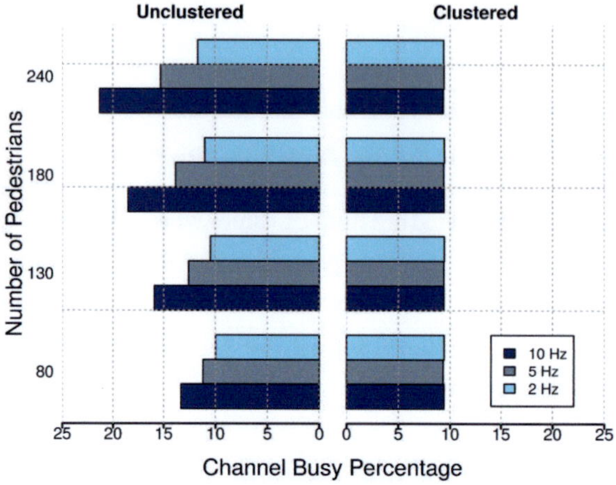

Figure 5.3: Channel Busy Percentage [SS20]

Figure 5.4 shows the comparison of the total number of pedestrian-generated beacons. The total number of beacons increases quickly with the number of pedestrians and the periodicity of pedestrian-generated safety messages in an un-clustered configuration. This is conforming with the findings of CBP for the same configurations. For clustered configuration, the number of beacons does not rise as quickly as the un-clustered configurations. The number of beacons is brought down in the configurations with 5 Hz and 10 Hz periodicity. However, for the 2 Hz periodicity configurations, the number of beacons is higher than their un-clustered counterparts. This is because, in the clustered configuration, CHs always broadcast the CSMs with 10 Hz periodicity.

Figure 5.5 shows the comparison of the average power consumption per second for the pedestrian devices between the clustered and un-clustered configurations. The power consumption increases linearly with the beacon periodicity in the case of un-clustered configurations. The power consumption per pedestrian device is lower for the clustered configurations and remains between 25-32 *battery units* per second. Also, the power savings increase with the beacon periodicity.

Figure 5.6 shows the comparison of the maximum number of *battery units* consumed by any pedestrian device during the given configuration. The pedestrian devices consume an equal amount of *battery units* for the given periodicity in an un-clustered configuration. This is expected as the power consumption of the

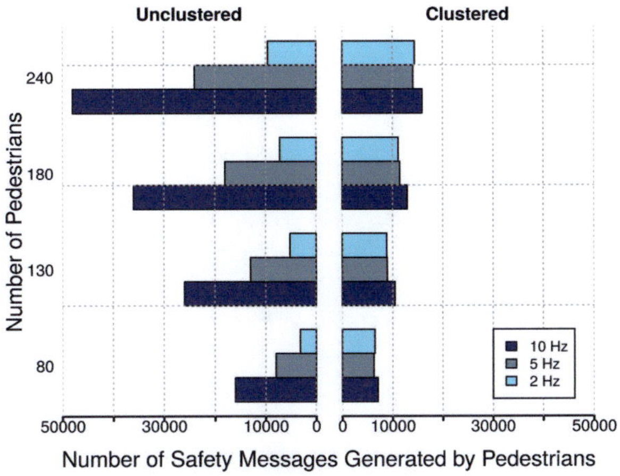

Figure 5.4: Total number of pedestrian beacons [SS20]

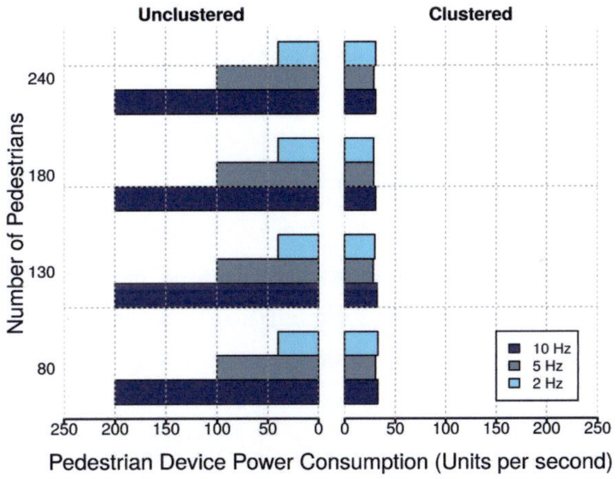

Figure 5.5: Average power consumption of pedestrian devices [SS20]

pedestrian devices is dependent on the beaconing periodicity and they all transmit with the same periodicity. However, for the clustered scenario, the consumed power depends on the role (CM vs. CH) that a pedestrian device may perform. The maximum amount of power consumed is less in a clustered configuration for 10 Hz periodicity. This is because even if CHs transmit CSMs with 10 Hz, the node may not continuously perform the CH role and may switch to CM role if a better CH comes along. For 5 Hz periodicity, the maximum power consumption is slightly higher in a clustered configuration. For the 2 Hz periodicity configuration, the difference between power consumption is high. This is because the CHs always transmit the CSMs with 10 Hz periodicity and lead to much higher power consumption compared to the 2 Hz configuration.

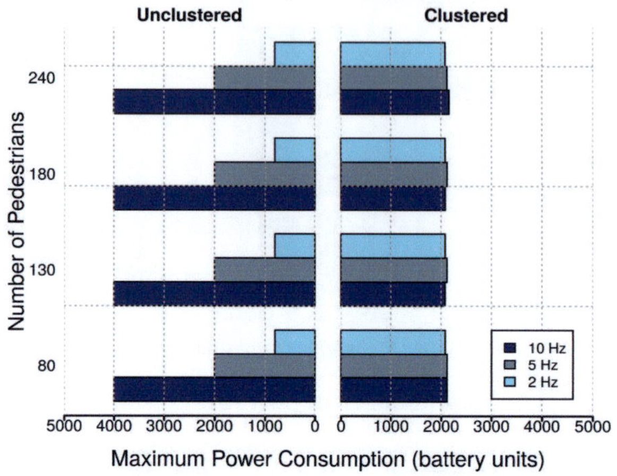

Figure 5.6: Maximum power consumption of pedestrian devices [SS20]

5.4 Discussion

This work demonstrates that the network congestion caused by the pedestrian-generated safety messages can be mitigated using clustering of pedestrians. It conceptualizes a novel mechanism of cooperative safety communication among pedestrians to minimize their impact on the broader collaborative V2X networks. This section discusses various aspects of the mechanism that may lay a foundation for further research.

5.4.1 Cluster parameters

MC-COCO4V2P primarily uses pedestrian movement patterns as the criteria for forming clusters. Although clustering has been extensively researched for VANETs, clustering of pedestrians for V2P communication is a novel concept and hence, opens up new possibilities for research. The pedestrian movement patterns differ significantly from that of the vehicles. This opens up new research possibilities for finding the optimum cluster configurations. Also, the clusters may be formed with not just pedestrians but also with other types of VRUs (such as bicyclists). It is also possible that clusters may consist of VRUs with varying mobility characteristics. The clustering mechanisms need to consider the impact of such scenarios on the cluster performance.

5.4.2 Channel modeling

The channel model is one of the necessary components for reliable development and verification of the new communication protocols. The clustering mechanisms in VANETs already lack adequate channel models due to the dynamic VANET environment [CFR+17]. Clustering mechanisms in V2P, such as MC-COCO4V2P, may increase the complexity of channel models further as they may need to support varying power levels, heterogeneous node characteristics, and varying communication device capabilities. This leads to the requirement of accurate channel models that can support such complex communication scenarios for reliable validation.

5.4.3 Communication technology for pedestrian collaboration

The exchange of cluster control messages among the pedestrians needs to be carried over a different medium than the safety communication. MC-COCO4V2P uses the SCH channel of DSRC architecture to ensure the separate medium. However, this may also be achieved using any other communication technology which can support ad-hoc communication. As smartphones are equipped with multiple communication technologies, this may be possible. However, pedestrians may not always explicitly enable cooperative communication and hence, the collaboration process needs to be autonomous and independent of the user input.

5.4.4 Infrastructure-based V2P approaches

V2P communication-based safety system is a relatively new concept. There has been no support for V2P communication in the infrastructure-based cellular network systems. However, this is evolving with the development of the new 5G standards. Multi-access Edge Computing (MEC) is an upcoming 5G technology [ETS18a] which enables the exchange of V2P safety messages. As MEC nodes can be deployed as part of the infrastructure, they may support the V2P communication with enough computing and power resources. They may also support communication using heterogeneous communication technologies. This may improve the ubiquity and reliability of V2P communication. However, the scalability of MEC-assisted V2P systems is so far unseen. Also, MEC applications are meant to provide services in a small geographical area (such as an intersection) and hence, the deployment of multiple MECs to cover a large geographical area may not be economical.

5.4.5 Reliability of intra-cluster and inter-cluster communication

The intra-cluster communication for pedestrian collaboration uses low transmit power so that it reaches only a few meters and does not create any interference with the ongoing safety communication. However, this can also have an opposite effect i.e. the high-powered safety transmission by vehicles can create interference for the intra-cluster communication and affects its reliability. Further, high pedestrian density can lead to high network load on the intra-cluster communication medium. These possibilities show that the parameters affecting the intra-cluster communication need to be studied further. This may help improve the reliability of intra-cluster communication and open few possibilities for further optimization.

Currently, MC-COCO4V2P does not support inter-cluster communication (among CHs). However, inter-cluster communication is another aspect that may be researched further to improve network congestion, forwarding of CSMs to extend their range, etc. In such cases, the reliability of inter-cluster communication also becomes an important aspect.

5.4.6 Accuracy of the sensor data

Based on the multiple V2P efforts, it has become evident that the smartphone will most likely be the pedestrian device of choice. Although smartphones are equipped with multiple sensors, these sensors may provide noisy data. Such noisy data can affect the accuracy of the clustering mechanism as well as the V2P safety communication. A good clustering mechanism needs to consider the possibility of such inaccuracies and operate accordingly.

5.4.7 Security and privacy

As pointed out before, cooperative safety communication needs to be autonomous requiring minimal user input. However, this demands that the pedestrian devices must be able to trust each other. This can be achieved by exchanging certificates that are issued by trustworthy entities, such as infrastructure operators [WSG+13].

6 On-demand Quality of Service for Crucial Communication

This chapter describes the mechanism that helps provide Quality-of-Service (QoS) to the crucial V2P communication. It describes the system model of the proposed mechanism, evaluation details, and results of the evaluation. The proposed solution has been submitted to IEEE Networking Letters [SSS20].

6.1 Introduction

When a pair of a vehicle and a VRU predicts that they are on the course of collision, they require reliability in their communication due to its crucial nature. As discussed in section 3.4.3, a mechanism to provide high reliability for crucial communication is required.

This chapter introduces a DSRC/WAVE architecture-based cooperative reliability mechanism that provides an 'On-demand QoS for crucial communication'. This mechanism focuses on the pair of vehicle and pedestrian that are on the verge of collision in a given area. It uses a special message that is broadcast by the pair to the surrounding nodes (vehicles, VRUs, etc.). This message notifies the surrounding nodes about the occurrence of crucial communication. Upon reception, the nodes may choose to lower the priority of their safety messages temporarily which results in better channel access for the critical pair. This results in the improved reliability of the crucial communication.

6.2 System Model

This section describes the system model of the proposed mechanism.

6.2.1 Concept

As discussed in section 2.3.1.2, the WAVE architecture supports an EDCA-based QoS mechanism for various types of V2X messages. The safety messages are always transmitted with the highest EDCA priority i.e. AC_VO. Although these messages are critical for safety, the crucial communication between the vehicle and the pedestrian requires even higher priority. However, allowing higher priority than AC_VO is currently not supported in the WAVE architecture. Hence, the higher priority for the crucial communication may be provided by temporarily reducing the priority of safety messages.

This work proposes an on-demand QoS mechanism that relies on sending a Priority Request Message (PRM) to the surrounding nodes. The PRM notifies the surrounding nodes about the ongoing crucial communication. When the surrounding nodes receive the PRM, they may choose to lower the periodicity of the safety messages. This would allow better channel access to the crucial pair which in turn would provide better reliability for the crucial communication. The PRM is transmitted by the vehicle-pedestrian pair on the same MAC channel as that of safety messages.

When the surrounding nodes receive the PRM from the vehicle-pedestrian pair, it is possible that some of the nodes may have ongoing crucial communication. For this purpose, this work assumes that not all the surrounding nodes would lower the priority of their transmission upon reception of the PRM. Also, in order to improve the reliability of delivery of PRM, this work proposes that the PRM be transmitted twice consecutively within the same time-slot (please refer to the figure 2.4 for time-slot division).

PRM is based on the WAVE Service Advertisement (WSA) message of the WAVE architecture and maintains backward compatibility.

6.2.2 Priority Request Message

PRM is a specially designed WSA message that notifies the surrounding nodes about the crucial communication. WSA packet format and its components, as defined by the WAVE architecture, are discussed in the section 2.3.1.3 and also, are shown in figures 2.11, 2.12, 2.13, 2.14, and 2.15. These elements are used as a baseline to decide the PRM content.

PRM WSA indicates the presence of the *WAVE Information Element* using the field *WAVE Information Element Extension.* PRM WSA defines a new *WAVE Information Element* with *WAVE Element ID* value 24. This value of the *WAVE*

Element ID is currently not allocated for any purpose and can be used to propose new applications. This new *WAVE Information Element* is 10 bytes in length and contains the following fields:

- *WAVE Element ID* - 1 byte (value set to 24)
- Length - 1 byte (value set to 8 bytes)
- Latitude - 4 bytes
- Longitude - 4 bytes

The *Latitude* and *Longitude* fields specify the location of the transmitter node.

Figure 6.1 illustrates the contents of *WSA Header Option Indicator* field of the PMR WSA. The *WAVE Information Element Extension* field of the *WSA Header Option Indicator* is set to 1 to indicate its presence and all other fields are set to 0. This tells the WAVE-capable device that a *WAVE Information Element* is present in the received WSA.

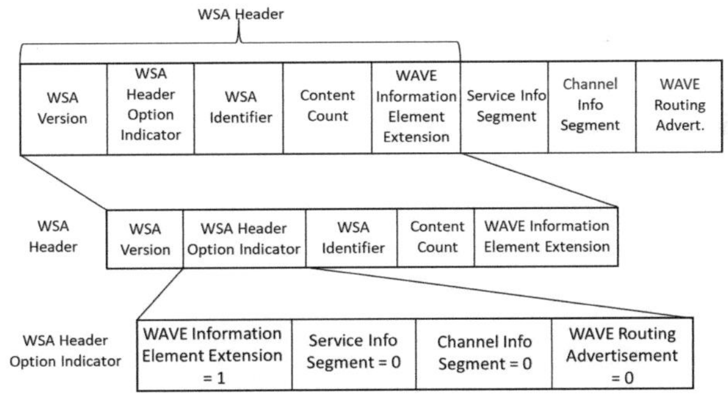

Figure 6.1: WSA Header Option Indicator [SSS20]

Figure 6.2 shows the *WAVE Information Element Extension* field of the PMR WSA header. It indicates the presence of 1 *WAVE Information Element.* The *WAVE Element Id* of this *WAVE Information Element* is set to 24 which indicates that it is a notification of the presence of crucial communication. If this *WAVE Element Id* is known to the receiving node, it may process it to take appropriate action. If the *WAVE Element Id* is not known, the node may discard the WSA.

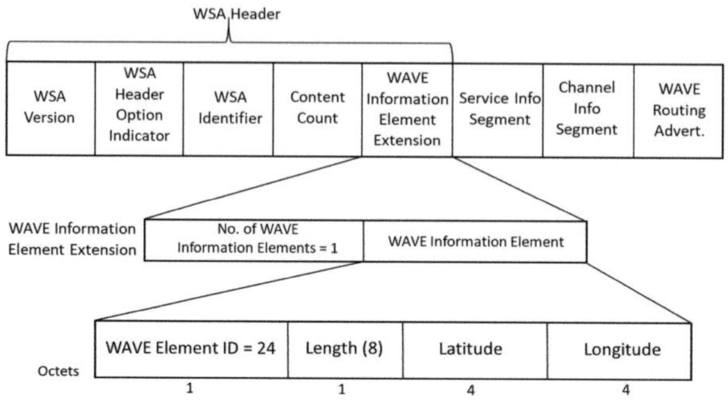

Figure 6.2: WAVE Information Element [SSS20]

6.2.3 Processing of Priority Request Message

6.2.3.1 Priority Request Message on Transmitter Side

The process involved in the generation of WSA is described in section 2.3.1.3. The process of generation of PRM, being a WSA, is similar to this process. The V2P collision warning app sends a *Provider service request* to WAVE Management Entity (WME) for the PRM generation. WME forwards the request to the *data plane* of the WSMP entity which then constructs the PRM WSA and sends it to lower layers for transmission.

6.2.3.2 Priority Request Message on Receiver Side

The processing of a received PRM WSA is similar to the procedure described in section 2.3.1.3. When the node receives the PRM WSA, it is forwarded to the WME entity. The WME entity verifies its validity and forwards it to the V2P collision warning app. The V2P collision warning application then decides to lower the priority of the safety messages. If the application already has another critical communication in progress it may decide not to lower the priority.

Also, upon reception of the PRM WSA, a certain proportion of surrounding nodes may agree to lower the priority. However, the priority may be lowered for an extended amount of time and not just for the immediate safety message. This may improve the overall reliability of the crucial communication further.

As discussed in the previous section, PRM is a specially designed WSA to notify surrounding vehicles about the presence of crucial communication. If the receiver node does not know the newly designated *WAVE Element Id*, it may discard the PRM WSA.

6.3 Performance Evaluation

This section discusses the evaluation of the proposed mechanism to verify the improvement in the crucial communication between the critical vehicle-pedestrian pair.

6.3.1 Scenario

The on-demand QoS mechanism is implemented and evaluated using the DSRC/WAVE protocol stack in OMNeT++, Veins, and SUMO. The new PRM WSA definition and the processing of PRM are implemented according to the described system model.

An urban intersection scenario, which may see a high footfall of pedestrians, is considered evaluation. The intersection sees the vehicle and pedestrian traffic and also, a vehicle-pedestrian pair that is on the verge of collision. Figure 6.3 depicts the intersection scenario used for evaluation. The intersection is formed by two orthogonal road segments that are 500 m long. There are a total of 4 footpaths of 200 m in length each along with the road segments. The vehicles are inserted at both the ends of each road segment and they travel in a straight direction to the other end. After reaching the other end, they exit the simulation. The pedestrians are inserted at the far end of the footpath and they travel towards the intersection. The pedestrians exit the simulation at the intersection. There is also a pair of a vehicle and a pedestrian that is assumed to be on the verge of collision under the pre-crash scenario 4.1b. The scenario contains obstacles that simulate the buildings in the urban environment.

The speed of the vehicles and the pedestrians is set at 13.89 m/s (50 km/h) and 1.3 m/s, respectively which is a typical urban traveling speed for these road entities. All the vehicles and pedestrians are tuned to CCH in continuous access mode. The vehicles broadcast the safety messages on CCH with 10 Hz periodicity. The pedestrians broadcast the safety messages on CCH with various pre-set periodicities. The pedestrians' safety message periodicity is chosen from 2, 5, and 10 Hz. The pair of vehicle-pedestrian involved in the crash transmit the safety messages with 10 Hz periodicity.

The system model assumes that not all nodes in the scenario will agree to lower the priority. This assumption has been modeled in the scenario. The scenario involves separate configurations where 25%, 50%, and 75% of the surrounding nodes agree to lower their EDCA priority upon reception of the PRM WSA. This allows studying the performance of the proposed mechanism under the different

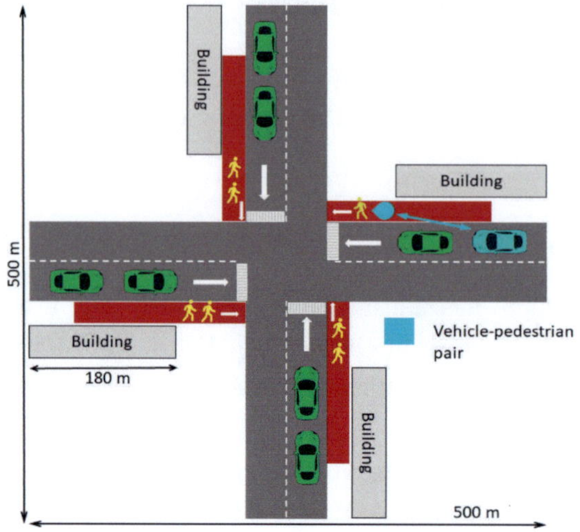

Figure 6.3: Scenario [SSS20]

proportions of the agreement by surrounding nodes.

The system model proposes that the surrounding nodes lower the priority for an extended amount of time. This requires that different lengths of time, during which the priority of safety messages would be lowered, be assessed for the highest effectiveness. For this purpose, the scenario models 3 different time windows of length 1 s, 3 s, and 5 s, respectively. This allows evaluating the performance of the critical communication during the specified window.

6.3.2 Simulation

As discussed in section 6.3.1, the on-demand QoS mechanism is implemented and evaluated using the OMNeT++, Veins, and SUMO simulation framework.

For the simulation purpose, the number of vehicles in the scenario is set to 150. The number of pedestrians is changed between 150 and 300. This allows studying the performance of PRM WSA under different network conditions and different pedestrian densities. Both the vehicle and pedestrian nodes are inserted at 1 s interval. All of the nodes in the simulation broadcast the messages with 20 mW power. This allows the communication range of up to 400 m. Table 6.1 shows the details of simulation parameters.

The pedestrian, from the pair, transmits the PRM WSA at the simulation time of 30 s. The different window sizes are then applied at this time to evaluate the performance of crucial communication. For example, the window size of 1 s is applied to measure the performance of crucial communication between simulation time 30.0 s to 31.0 s.

Simulation Parameters	Value
Road length	500 m
Road layout	Two-way, two-lanes
No. of vehicles	150
Max vehicle speed	13.89 m/s = 50 km/h
No. of Pedestrians	150, 300
Max pedestrian speed	1.3 m/s
Vehicle beacon periodicity	10 Hz
Pedestrian beacon periodicity	2,5,10 Hz
Vehicle Tx Power	20 mW
Pedestrian Tx power (BSM)	20 mW
Pedestrian Tx power (PRM WSA)	20 mW
BSM size	135 bytes
Priority Request Message size	121 bytes
Nodes in agreement	25%, 50%, 75%
Window size	1 s, 3 s, 5 s

Table 6.1: Simulation parameters [SSS20]

To precisely control the percentage of nodes in agreement, nodes are pre-selected for lowering the priority of safety message transmission. These nodes lower the priority of safety messages during the selected window size.

The simulation environment employs *TwoRayInterferenceModel* and *SimpleObstacleShadowing* models to simulate the signal propagation characteristics in the urban environment. The simulation warm-up period is set to 30 s which is the same as the simulation time of the beginning of PRM WSA transmission. This allows the intersection to attain the populated network state. The traffic light in both directions has 30 s and 5 s of green and yellow phases, respectively. The simulation run time is set according to the window size. So, the simulation lengths for window sizes 1 s, 3 s, and 5 s are set to 31 s, 33 s, and 35 s, respectively. The data for each configuration is gathered over three independent runs using different seeds.

6.3.3 Evaluation

The on-demand QoS mechanism aims to improve the reliability of crucial communication between the vehicle-pedestrian pair. Following metrics are used to assess the impact of the on-demand QoS mechanism.

1. Crucial Beacon Delivery Ratio

As discussed in section 3.5.3.2, the crucial Beacon Delivery Ratio (B_cDR) can provide insights into the QoS for crucial communication between the pair. B_cDR at the vehicle and pedestrian side is given by equations 6.1 and 6.2, respectively.

$$B_cDR_{(V)} = \frac{num_{\text{received(Vehicle)}}}{num_{\text{sent(Pedestrian)}}} \tag{6.1}$$

where:

$num_{\text{received(Vehicle)}}$ = number of beacons received by vehicle node
$num_{\text{sent(Pedestrian)}}$ = number of beacons sent by pedestrian node

$$B_cDR_{(P)} = \frac{num_{\text{received(Pedestrian)}}}{num_{\text{sent(Vehicle)}}} \tag{6.2}$$

where:

$num_{\text{received(Pedestrian)}}$ = number of beacons received by pedestrian node
$num_{\text{sent(Vehicle)}}$ = number of beacons sent by vehicle node

Average B_cDR for the pair for any configuration ($B_cDR_{(Avg)}$) is calculated by taking the average of $B_cDR_{(V)}$ and $B_cDR_{(P)}$ for that configuration. $B_cDR_{(Avg)}$ calculation is shown in equation 6.3.

$$B_cDR_{(Avg)} = \frac{B_cDR_{(V)} + B_cDR_{(P)}}{2} \tag{6.3}$$

2. Channel Busy Percentage

Channel Busy Percentage (CBP) can be used to gauge the channel load of the V2X network. Although CBP is not the direct indicator of crucial communication, it can be used to validate the trends in the performance of the crucial communication. CBP is determined using the equation 6.4. CBP for every node (pedestrians and vehicles) is calculated and then the mean value is used to gauge the channel load.

$$CBP = \frac{t_{\text{ChBusy}}}{t_{\text{CBP}}} \times 100\% \tag{6.4}$$

where:

t_{ChBusy} = duration of the time when the channel was busy
t_{CBP} = measurement window

Figures 6.4, 6.5, and 6.6 show the comparison of $B_cDR_{(Avg)}$ results for the 25%, 50%, and 75% node agreement configurations, respectively. The figures show that with or without WSA, $B_cDR_{(Avg)}$ decreases with the increase in the beaconing periodicity. This is expected because the channel load increases with the pedestrian beaconing periodicity as shown by figure 6.8.

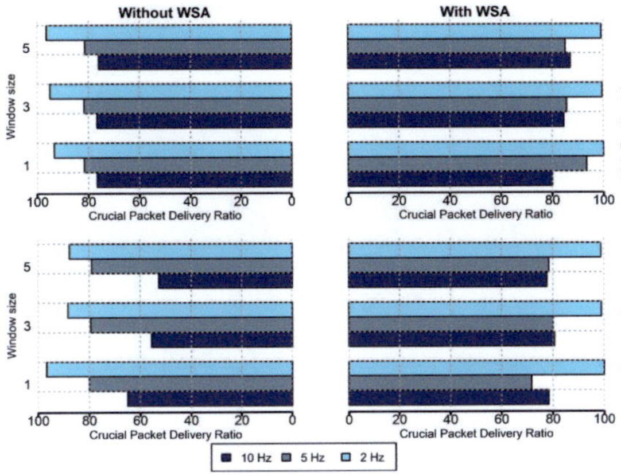

Figure 6.4: Crucial beacon delivery ratio for 25% node agreement [SSS20]

The $B_cDR_{(Avg)}$ results for 25% node agreement with 300 pedestrians show that $B_cDR_{(Avg)}$ is improved for 10 Hz periodicity for all window sizes. Also, $B_cDR_{(Avg)}$ is improved for all windows sizes and pedestrian configurations with 2 Hz periodicity. For 300 pedestrians with 5 Hz periodicity, $B_cDR_{(Avg)}$ decreases for window size of 1 s and 5 s. To investigate this issue further, simulations were run only with the vehicle-pedestrian pair (without any other nodes being present). In such case, the $B_cDR_{(Avg)}$ was 100%. It is possible that for this particular configuration, the pair may be suffering through a hidden node problem. The hidden node problem is known to occur in the system using CSMA/CA for channel access.

The results for 50% node agreement (figure 6.5) show that the $B_cDR_{(Avg)}$ is improved for all configurations except for the case of 150 pedestrians with 2 Hz periodicity where they are nearly equal.

The results for 75% node agreement (figure 6.6) show that the $B_cDR_{(Avg)}$ is improved for 5 Hz and 10 Hz periodicity under all window sizes and pedestrian configurations. Under the 75% node agreement 2 Hz periodicity configuration, $B_cDR_{(Avg)}$ is nearly equal for under the window sizes 3 s and 5 s. It has worsened for 2 Hz periodicity with the window size 1 s. The $B_cDR_{(Avg)}$ for the 75% node agreement with 150 pedestrians, 2 Hz periodicity, and window size 1 s without WSA is 93.33% ± 11.55 while that with WSA is 90.0% ± 5.0. This shows that the $B_cDR_{(Avg)}$ is within the range of the state of the art (without WSA) but it also assures packet delivery more consistently. To investigate the inconsistency

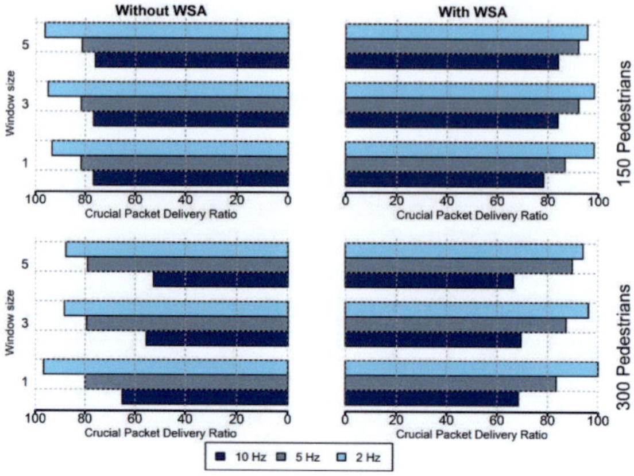

Figure 6.5: Crucial beacon delivery ratio for 50% node agreement [SSS20]

with 300 pedestrians configuration, $B_cDR_{(Avg)}$ was measured without any other nodes being present. For such scenario, the $B_cDR_{(Avg)}$ value was 100%. This configuration may also be suffering from the hidden node problem.

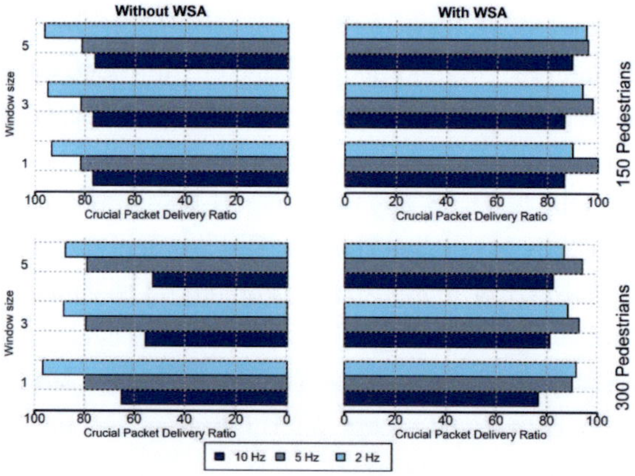

Figure 6.6: Crucial beacon delivery ratio for 75% node agreement [SSS20]

Figure 6.7 shows the percentage improvement in the crucial beacon delivery ratio ($B_cDR_{(Avg)}$). The figure confirms the improvements and anomalies discussed above. Also, it can be seen that the improvement is highest for the configurations with 10 Hz periodicity. 25% node agreement configuration shows the highest improvements in 2 Hz periodicity configurations while 75% node agreement shows the highest improvement in 5 Hz and 10 Hz configurations. 50% node agreement configurations show consistently small improvements across all configurations except the one with 150 pedestrians and 2 Hz periodicity.

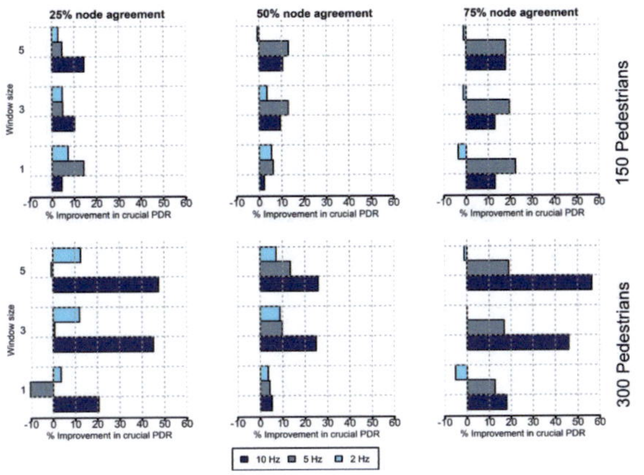

Figure 6.7: Improvement in Crucial beacon delivery ratio [SSS20]

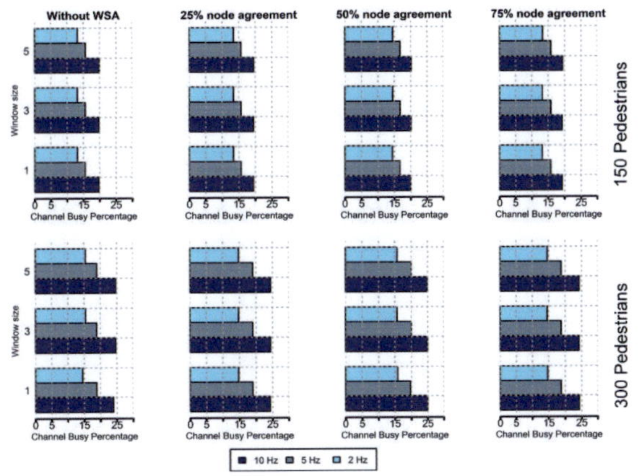

Figure 6.8: Channel Busy Percentage [SSS20]

Figure 6.8 shows the CBP of all configurations. It confirms that the network load increases with the pedestrian beaconing frequency as well as the number of pedestrians. Also, it shows that the CBP is unchanged for any given periodicity and number of pedestrians across all node agreements. This is expected as the two PRM WSAs and the lower priority messages are unlikely to affect the CBP significantly.

6.4 Discussion

This work demonstrates that the reliability of crucial communication between the pair of vehicle and pedestrian can be improved by indicating the presence of such crucial communication. The work proposes a concept that makes a further distinction between crucial communication and safety-critical communication. However, it is important to understand that not all vehicles may comply with the request due to the different needs of various ongoing applications on vehicles. The effects of such different levels of complying are shown using different percentages of node agreement. Further, the work assumes that vehicles may comply with uniform window sizes. However, it may happen that the vehicles comply with the request with different window sizes. The effects of such mixed window sizes need to be evaluated further.

The concept of an on-demand QoS may also be applied to other crucial communication scenarios in the V2X networks. For example, one of the common scenarios involves a vehicle that wants to make a left turn in the presence of oncoming traffic. The communication involved in the negotiation process may qualify as a crucial communication and may benefit from on-demand QoS. Such scenarios of crucial communication need to be identified and evaluated further. An alternative to the approach of lowering the *priority* temporarily upon request is to lower the *periodicity* of safety messages generated by vehicle and pedestrians upon request [SS19b]. However, this approach needs to be studied and developed further.

7 Summary

V2P communication systems can extend the capabilities of existing driver assistance systems of the vehicles. Although V2P systems can leverage the well-researched and standardized V2V network protocols, V2P systems face different challenges due to the differing characteristics of VRUs. These challenges of V2P systems have not been addressed by the current state-of-the-art V2X literature. This dissertation identifies these challenges and proposes solutions for them. This chapter provides a summary of these contributions, discusses the practical challenges, and lists future work.

7.1 Research Contributions

The dissertation discusses the basics of the V2P crash prevention system where it provides insights into the V2P communication characteristics and the V2P system components. It discusses different V2P system aspects which need to be considered for the efficient operation of the V2P systems. These aspects present three individual challenges which include V2P system design, network congestion control, and QoS for crucial V2P communication. This dissertation proposes individual solutions for each of these challenges.

The presented V2P design framework provides a systematic and holistic approach towards the design of V2P systems. The V2P design framework is based on 9 different categories of inputs. The framework is studied and evaluated by choosing different values of design inputs from different categories. The categories chosen for evaluation include pre-crash scenarios, mode of communication, different VRUs, and response time requirements. The simulation results show that the different values of design inputs affect the overall performance of the V2P system.

Moreover, this dissertation proposes MC-COCO4V2P, an energy-efficient clustering mechanism for pedestrians, which helps in network congestion mitigation caused by the pedestrian-generated safety messages. The mechanism involves the pedestrian devices communicating with each other in order to form clusters. These clusters are then used for coordinated transmission of safety messages on behalf of the pedestrians. Simulations show that the network load is mitigated when the MC-COCO4V2P mechanism is employed. MC-COCO4V2P represents a novel concept of collaborative communication among pedestrians in order to reduce the impact on broader V2X safety communication. It is technology-agnostic and can be implemented using any communication technology that allows ad-hoc communication with minimal user input.

The third challenge of QoS for crucial communication is addressed by providing an on-demand QoS mechanism. The on-demand QoS is a novel concept for broadcast-based V2X networks. It involves the transmission of special messages to make surrounding nodes aware of the crucial communication. The surrounding nodes then may decide to lower the priority of their safety messages. The mechanism also considers the possibility that not all nodes may decide to lower the priority of safety messages. This helps provide realistic feedback for the improvement in the reliability of crucial communication. The simulations show that the performance of crucial communication is improved in most of the configurations.

7.2 Practical Challenges

This section describes the underlying assumptions that this dissertation has made about the V2P systems. Practical challenges may arise if these assumptions are not made.

It assumes that every VRU device and vehicle is equipped with V2X communication technology (specifically, DSRC). Further, a smartphone may be a suitable VRU device. As V2P (and V2X in general) is currently in a nascent stage, it may take some time for smartphones and vehicles to be equipped with V2X technologies. This may pose deployment challenges in the intermediate time horizon.

The dissertation also assumes that smartphones (VRU devices) provide accurate context data, which can be used as part of safety message transmission. In real-life implementations, the context data provided by smartphones may be noisy and hence, may need to be processed with estimation algorithms (for example, Kalman filter).

Also, V2X systems are meant to extend the capabilities of existing driver assistance systems. The simulation framework used by this dissertation primarily provides the V2X communication capabilities in stand-alone mode. In order to have more realistic studies of pre-crash scenarios, it may be helpful to have a simulated driver assistance module (for example, LIDAR) working in tandem with the vehicular V2X system.

7.3 Future Work

The V2P system design framework currently provides the categories for a V2P system based on V2X communications. It is possible to extend it further with the addition of more categories, such as the role of vehicles, the role of Geographical Information Systems (GIS), etc. However, this may require further studies of the VRU crash statistics at a more granular level and the identification of appropriate scenarios for targeted VRU groups. Also, the last few years have seen the rise of new categories of VRUs, such as e-scooter riders. Mobility characteristics of such VRUs have to be studied and current V2P systems have to be evaluated for such new categories.

The network congestion control mechanism proposed in this dissertation currently relies on direct communication, based on DSRC, among the pedestrians. However, it would be possible to achieve clustering of pedestrians through other adhoc technologies, such as Wi-Fi direct, or through upcoming 5G Device-2-Device (D2D) technologies. It is also possible to delegate the cluster management role to cellular infrastructure nodes. However, such mechanisms need to be designed and evaluated to determine the promising candidates. The further possibilities with pedestrian clustering and the hints for future work can be found under the *Discussion* section of the chapter *Network Congestion Control in V2P Networks.*

The concept of On-demand QoS for crucial communication is introduced in this dissertation. The further possibilities of development of On-demand QoS and the hints for future work have been discussed under the *Discussion* section of the chapter *On-demand Quality of Service for Crucial Communication.* Further, this concept may be broadly applied to the other V2X systems. A new message may be developed for the C-ITS system (European V2X system) that may be used for on-demand QoS. Similarly, for C-V2X technology, new resource selection algorithms may be developed which may adapt to the on-demand QoS request.

Bibliography

[3GP19a] 3GPP: *3GPP TS 23.285 V16.2.0 - Architecture enhancements for V2X services (Release 16)*. 2019

[3GP19b] 3GPP: *3GPP TS 36.300 V15.8.0: Overall description; Stage 2 (Release 15)*. 2019

[AKY17] Artail, Hassan ; Khalifeh, Kareem ; Yahfoufi, Mohamad: Avoiding Car-Pedestrian Collisions Using a VANET to Cellular Communication Framework. In: *13th International Wireless Communications and Mobile Computing Conference*, IEEE, 2017

[AMS+14] Anaya, Jose J. ; Merdrignac, Pierre ; Shagdar, Oyunchimeg ; Nashashibi, Fawzi ; Naranjo, Jose E.: Vehicle to Pedestrian Communications for Protection of Vulnerable Road Users. In: *Intelligent Vehicles Symposium (IV)* IEEE, 2014

[AOW12] Almalag, Mohammad S. ; Olariu, Stephan ; Weigle, Michele C.: TDMA Cluster based MAC for VANETs (TC-MAC). In: *IEEE International Symposium on a World of Wireless, Mobile and Multimedia Networks (WoWMoM)*, IEEE, 2012

[APB12] Ajaltouni, Hikmat E. ; Pazzi, Richard W. ; Boukerche, Azzedine: An Efficient QoS MAC for IEEE 802.11p Over Cognitive Multichannel Vehicular Networks. In: *IEEE International Conference on Communications (ICC)*, IEEE, 2012

[ARI12] ARIB: *ARIB STD-T109: 700 MHz Band Intelligent Transport Systems*. 2012

[AS13] Allouche, Yair ; Segal, Michael: A Cluster Based Beaconing Process for VANET. In: *Proceeding of the 10th ACM International Workshop on Vehicular Inter-networking, Systems, and Applications*, ACM, 2013

[ATG+15] Anaya, José J. ; Talavera, Edgar ; Giménez, David ; Gómez, Nuria ; Jiménez, Felipe ; Naranjo, José E.: Vulnerable Road Users Detection using V2X Communications. In: *18th International Conference on Intelligent Transportation Systems*, IEEE, 2015

[Aut17] Authority, Tampa Hillsborough E.: *Tampa Connected Vehicle Pilot*. https://www.tampacvpilot.com/. Version: 2017

[BHLS18] Boulila, Narjes ; Hadded, Mohamed ; Laouiti, Anis ; Saidane, Leila A.: QCH MAC: A Qos-Aware Centralized Hybrid MAC Protocol for Vehicular Ad Hoc NETworks. In: *IEEE 32nd International Conference on Advanced Information Networking and Applications AINA*, IEEE, may 2018

[BKR13] Bansal, Gaurav ; Kenney, John B. ; Rohrs, Charles E.: LIMERIC: A Linear Adaptive Message Rate Algorithm for DSRC Congestion Control. In: *IEEE Transactions on Vehicular Technology* 62 (2013), S. 4182 – 4197

[BLKP13] Bansal, Gaurav ; Lu, Hongsheng ; Kenney, John B. ; Poellabauer, Christian: EMBARC: Error Model based Adaptive Rate Control for Vehicle-to-Vehicle Communications. In: *Proceeding of the tenth ACM international workshop on Vehicular internetworking, systems, and applications*, ACM, 2013

[BSN14] Bagheri, Mehrdad ; Siekkinen, Matti ; Nurminen, Jukka K.: Cellular-based Vehicle to Pedestrian (V2P) Adaptive Communication for Collision Avoidance. In: *International Conference on Connected Vehicles and Expo (ICCVE)*, IEEE, 2014

[CFM+13] Carpenter, Michael G. ; Feldmann, M. ; Moury, M T. ; Skvarce, Jeffrey R. ; Struck, Matthias ; Zwicky, Timothy D. ; Kiger, Steven M.: Objective Tests and for and Forward-Looking Pedestrian and Crash Avoidance/Mitigation - Systems Annual Report / National Highway Traffic Safety Administration. 2013 (DOT HS 811 793). – Forschungsbericht

[CFR+17] Cooper, Craig ; Franklin, Daniel ; Ros, Montserrat ; Safaei, Farzad ; Abolhasan, Mehran: A Comparative Survey of VANET Clustering Techniques. In: *IEEE Communications Surveys & Tutorials* 19 (2017), Nr. 1, S. 657–681

[Cro12] CRONIN, Brian: *Vehicle Based Data and Availability.* https://www.its.dot.gov/itspac/october2012/PDF/data_availability.pdf. Version: 2012

[CYD16] CHANG, Che-Yu ; YEN, Hsu-Chun ; DENG, Der-Jiunn: V2V QoS Guaranteed Channel Access in IEEE 802.11p VANETs. In: *IEEE Transactions on Dependable and Secure Computing* 13 (2016), jan, Nr. 1, S. 5–17. http://dx.doi.org/10.1109/tdsc.2015.2399912. – DOI 10.1109/tdsc.2015.2399912

[DGD12] DJAHEL, Soufiene ; GHAMRI-DOUDANE, Yacine: A Robust Congestion Control Scheme for Fast and Reliable Dissemination of Safety Messages in VANETs. In: *IEEE Wireless Communications and Networking Conference: Mobile and Wireless Networks*, IEEE, 2012

[DSCP14] DHONDGE, Kaustubh ; SONG, Sejun ; CHOI, Baek-Young ; PARK, Hyungbae: WiFiHonk: Smartphone-based Beacon Stuffed WiFi Car2X-Communication System for Vulnerable Road User Safety. In: *79th Vehicular Technology Conference (VTC)* IEEE, 2014

[ELPM16] EGEA-LOPEZ, Esteban ; PAVON-MARINO, Pablo: Distributed and Fair Beaconing Rate Adaptation for Congestion Control in Vehicular Networks. In: *IEEE Transactions on Mobile Computing* 15 (2016), S. 3028 – 3041

[Eng13] ENGEL, Sebastian: Car2Pedestrian: Protection of vulnerable road users using smartphones. In: *17th International Forum on Advanced Microsystems for Automotive Applications*, 2013

[ETS09] ETSI: *ETSI TR 102 638 V1.1.1 :Intelligent Transport Systems (ITS); Vehicular Communications; Basic Set of Applications; Definitions.* 2009

[ETS10] ETSI: *ETSI EN 302 665 V1.1.1: Intelligent Transport Systems Communications Architecture.* 2010

[ETS11] ETSI: *ETSI TS 102 637-2: Intelligent Transport Systems (ITS); Vehicular Communications; Basic Set of Applications; Part 2: Specification of Cooperative Awareness Basic Service.* 2011

[ETS13a] ETSI: *ETSI EN 302 663 V1.2.1: Intelligent Transport Systems (ITS); Access layer specification for Intelligent Transport Systems operating in the 5 GHz frequency band.* 2013

[ETS13b] ETSI: *ETSI TS 102 894-1 V1.1.1: Intelligent Transport Systems (ITS); Users and applications requirements; Part 1: Facility layer structure, functional requirements and specifications.* 2013

[ETS14a] ETSI: *ETSI EN 302 636-3 V1.1.2: Intelligent Transport Systems (ITS); Vehicular Communications; GeoNetworking; Part 3: Network Architecture.* 2014

[ETS14b] ETSI: *ETSI EN 302 637-3 V1.2.1: Intelligent Transport Systems (ITS); Vehicular Communications; Basic Set of Applications; Part 3: Specifications of Decentralized Environmental Notification Basic Service.* 2014

[ETS16] ETSI: *ETSI TS 103 248 V1.1.1: Intelligent Transport Systems (ITS); GeoNetworking; Port Numbers for the Basic Transport Protocol (BTP).* 2016

[ETS17] ETSI: *ETSI EN 302 636-5-1 V2.1.0: Intelligent Transport Systems (ITS); Vehicular Communications; GeoNetworking; Part 5: Transport Protocols; Sub-part 1: Basic Transport Protocol.* 2017

[ETS18a] ETSI: *ETSI GR MEC 022 V2.1.1: Multi-access Edge Computing (MEC); Study on MEC Support for V2X Use Cases.* 2018

[ETS18b] ETSI: *Intelligent Transport Systems (ITS); Decentralized Congestion Control Mechanisms for Intelligent Transport Systems operating in the 5 GHz range; Access layer part.* 2018

[ETS19] ETSI: *ETSI EN 302 663 V1.3.0: Intelligent Transport Systems (ITS); ITS-G5 Access layer specification for Intelligent Transport Systems operating in the 5 GHz frequency band.* 2019

[Eur19] EUROPEAN TELECOMMUNICATIONS STANDARDS INSTITUTE SPECIALIST TASK FORCE 565: *Specifications for the definition of the cooperative ITS Vulnerable Road Users (VRU) service.* `https://docbox.etsi.org/Workshop/2019/201903_STF565_VRU/STF565_Presentation_190306.pdf`. Version: 2019

[FSYN15] Fujikami, Shintaro ; Sumi, Takenori ; Yagiu, Riko ; Nagai, Yukimasa: Fast Device Discovery for Vehicle-to-Pedestrian Communication using Wireless LAN. In: *12th Annual IEEE Consumer Communications and Networking Conference (CCNC)* IEEE, 2015, S. 35–40

[GAR17] Goudarzi, Forough ; Al-Raweshidy, Hamed S.: Distributed Transmit Power Control for Beacons in VANET. In: *3rd International Conference on Vehicle Technology and Intelligent Transport Systems (VEHITS 2017)*, 2017

[GAS09] Goonewardene, R.T. ; Ali, F.H. ; Stipidis, E.: Robust mobility adaptive clustering scheme with support for geographic routing for vehicular ad hoc networks. In: *IET Intelligent Transport Systems* 3 (2009), Nr. 2, S. 148. http://dx.doi.org/10.1049/iet-its:20070052. – DOI 10.1049/iet–its:20070052

[GCNB01] Gomez, Javier ; Campbell, Andrew T. ; Naghshineh, Mahmoud ; Bisdikian, Chatschik: Conserving transmission power in wireless ad hoc networks. In: *Proceedings Ninth International Conference on Network Protocols*, IEEE, 2001

[GMW12] General Motors Co.: *GM Developing Wireless Pedestrian Detection Technology.* GM Corporate Newsroom, http://media.gm.com/media/us/en/gm/news.detail.html/content/Pages/news/us/en/2012/Jul/0726_pedestrian.html. http://media.gm.com/media/us/en/gm/news.detail.html/content/Pages/news/us/en/2012/Jul/0726_pedestrian.html. Version: 2012

[Gro09] Group, BMW: *BMW Car-2-X Communication improves pedestrian safety.* BMW-Blog, http://www.bmwblog.com/2009/04/22/bmw-car-2-x-communication-improves-pedestrian-safety/, 2009

[Gua18] Guardian News: *Uber dashcam footage shows lead up to fatal self-driving crash.* https://www.youtube.com/watch?v=RASBcc4yOOo. Version: 2018

[GWG07] Gunter, Yvonne ; Wiegel, Bernhard ; Grossmann, Hans P.: Cluster-based Medium Access Scheme for VANETs. In: *IEEE Intelligent Transportation Systems Conference*, IEEE, 2007

[HC17] Ho, Ping-Fan ; Chen, Jyh-Cheng: WiSafe: Wi-Fi Pedestrian Collision Avoidance System. In: *IEEE Transactions on Vehicular Technology* 66 (2017), Nr. 6, S. 4564–4578

[HCT+16] Huang, Kuang-Shih ; Chiu, Po-Jui ; Tsai, Hsin-Mu ; Kuo, Chih-Chung ; Lee, Hui-Yu ; Wang, Yu-Chiang F.: RedEye: Preventing Collisions Caused by Red-Light-Running Scooters With Smartphones. In: *IEEE Transactions On Intelligent Transportation Systems*, IEEE, May 2016

[HFSK11] Huang, Ching-Ling ; Fallah, Yaser P. ; Sengupta, Raja ; Krishnan, Hariharan: Intervehicle Transmission Rate Control for Cooperative Active Safety System. In: *IEEE Transactions on Intelligent Transportation Systems* 12 (2011), S. 645 – 658

[HGAOM16] Hussein, Ahmed ; Garcia, Fernando ; Armingol, Jose M. ; Olaverri-Monreal, Cristina: P2V and V2P Communication for Pedestrian Warning on the basis of Autonomous Vehicles. In: *IEEE 19th International Conference on Intelligent Transportation Systems (ITSC)*, IEEE, 2016

[HHT11] Hsu, Chih-Wei ; Hsu, Chung-Hsien ; Tseng, Huei-Ru: MAC Channel Congestion Control Mechanism in IEEE 802.11p/WAVE Vehicle Networks. In: *IEEE Vehicular Technology Conference (VTC Fall)*, IEEE, 2011

[HJPI15] Hernandez-Jayo, Unai ; Perez, Jagoba ; Iglesia, Idoia de-la: Poster: Wearable Warning System for Improving Cyclists Safety in the Scope of Cooperative Systems. In: *IEEE Vehicular Networking Conference (VNC)*, IEEE, 2015

[HKDS18] Heinovski, Julian ; Klingler, Florian ; Dressler, Falko ; Sommer, Christoph: A simulative analysis of the performance of IEEE 802.11p and ARIB STD-T109. In: *Computer Communications* 122 (2018), jun, S. 84–92. `http://dx.doi.org/10.1016/j.comcom.2018.03.016`. – DOI 10.1016/j.comcom.2018.03.016

[IEE16a] IEEE: *IEEE Standard for Wireless Access in Vehicular Environments (WAVE) - Multi-Channel Operation: Std 1609.4^{TM}-2016.* 2016

[IEE16b] IEEE: *IEEE Standard for Wireless Access in Vehicular Environments (WAVE) - Networking Services:Std 1609.3^TM-2016.* 2016

[IEE16c] IEEE: *IEEE Std 1609.12^TM-2016: Standard for Wireless Access in Vehicular Environments (WAVE) - Identifier Allocations.* 2016

[IEE16d] IEEE Computer Society: *IEEE Std 802.11^TM-2016: IEEE Standard for Information technology - Telecommunications and information exchange between systems Local and metropolitan area networks - Specific requirements: Part 11: Wireless LAN Medium Access Control (MAC) and Physical Layer (PHY) Specifications.* 2016

[IEE16e] IEEE Vehicular Technology Society: *IEEE Std 1609.2^TM-2016: Standard for Wireless Access in Vehicular Environments - Security Services for Applications and Management Messages.* 2016

[IEE19] IEEE: *IEEE Guide for Wireless Access in Vehicular Environments (WAVE) Architecture: Std 1609.0^TM-2019.* 2019

[InD18] InDev: *In-depth Understanding of Accident Causation for Vulnerable Road Users.* `https://www.indev-project.eu/I`. Version: 2018

[Int14] International Transport Forum: *Road Safety Annual Report 2014 Summary.* `http://www.internationaltransportforum.org/pub/pdf/14IrtadReport.pdf`. Version: 2014

[IRU15] IRU: *Compass4D: Improving road safety and efficiency by sharing information wirelessly.* `https://www.iru.org/innovation/iru-projects/compass4d`. Version: 2015

[JLJ18] Joseph, Maan ; Liu, Xiaofeng ; Jaekel, Arunita: An Adaptive Power Level Control Algorithm for DSRC Congestion Control. In: *Proceedings of the 8th ACM Symposium on Design and Analysis of Intelligent Vehicular Networks and Applications*, ACM, 2018

[JMF99] Jain, A.K. ; Murty, M.N. ; Flynn, P.J.: Data Clustering: A Review. In: *Association for Computing Machinery* 31 (1999), Nr. 3, 264–323. `https://doi.org/10.1145/331499.331504`. – ISSN 0360–0300

[KA07] Kayis, Oemer ; Acarman, Tankut: Clustering Formation for Inter-Vehicle Communication. In: *IEEE Intelligent Transportation Systems Conference*, IEEE, 2007

[KA10] Klaus, David ; Alexander, Flach: CAR-2-X and Pedestrian Safety. In: *IEEE Vehicular Technology Magazine* (2010). – ISSN 1556–6072

[Kas17] Kassel: *VERONIKA - research project*. https://www.kassel.de/buerger/verkehr_und_mobilitaet/verkehrsprojekte/veronika.php. Version: 2017

[KBR11] Kenney, John B. ; Bansal, Gaurav ; Rohrs, Charles E.: LIMERIC: A Linear Message Rate Control Algorithm for Vehicular DSRC Systems. In: *Proceedings of the Eighth ACM international workshop on Vehicular inter-networking*, ACM, 2011

[KEBB12] Krajzewicz, Daniel ; Erdmann, Jakob ; Behrisch, Michael ; Bieker, Laura: Recent Development and Applications of SUMO - Simulation of Urban MObility. In: *International Journal On Advances in Systems and Measurements* 5 (2012), December, Nr. 3&4, S. 128–138

[KFHG19] Kelarestaghi, Kaveh B. ; Foruhandeh, Mahsa ; Heaslip, Kevin ; Gerdes, Ryan: *Survey on Vehicular Ad Hoc Networks and Its Access Technologies Security Vulnerabilities and Countermeasures*. 2019

[KHS+16] Kloiber, Bernhard ; Harri, Jerome ; Strang, Thomas ; Sand, Stephan ; Garcia, Cristina R.: Random Transmit Power Control for (DSRC) and its Application to Cooperative Safety. In: *IEEE Transactions on Dependable and Secure Computing* 13 (2016), jan, Nr. 1, S. 18–31. http://dx.doi.org/10.1109/tdsc.2015.2449845. – DOI 10.1109/tdsc.2015.2449845

[KKJ18] Kim, Jin-Woo ; Kim, Jae-Wan ; Jeon, Dong-Keun: A Co-operative Communication Protocol for QoS Provisioning in IEEE 802.11p/Wave Vehicular Networks. In: *Sensors* 18 (2018), oct, Nr. 11, S. 3622. http://dx.doi.org/10.3390/s18113622. – DOI 10.3390/s18113622

[KOV+14] KWAKKERNAAT, Maurice ; OPHELDERS, Frank ; VISSERS, John ; WILLEMSEN, Dehlia ; SUKUMAR, Premnaath: Cooperative Automated Emergency Braking For Improved Safety Beyond Sensor Line-Of-Sight And Field-Of-View. In: *FISITA 2014 World Automotive Congress*, 2014

[LBKW13] LEWANDOWSKI, Andreas ; BOECKER, Stefan ; KOESTER, Volker ; WIETFELD, Christian: Design and Performance Analysis of an IEEE 802.15.4 V2P Pedestrian Protection System. In: *5th International Symposium on Wireless Vehicular Communications*, IEEE, 2013

[LCC+16] LIN, Chi-Han ; CHEN, Yi-Ting ; CHEN, Jyun-Jie ; SHIH, Wen-Chan ; CHEN, Wen-Tsuen: pSafety: A Collision Prevention System for Pedestrians Using Smartphone. In: *84th Vehicular Technology Conference (VTC-Fall)*, IEEE, 2016

[LFYZ14] LU, Zhi-Qiang ; FENG, Yong ; YANG, Jie ; ZHENG, Xiong-Qi: An Improved Infrastructure Assisted V2P Protocol for VANETs. In: *10th International Conference on Wireless Communications, Networking and Mobile Computing (WiCOM)* IET, 2014

[LK16] LEE, Sungwon ; KIM, Dongkyun: An Energy Efficient Vehicle to Pedestrian Communication Method for Safety Applications. In: *Wireless Personal Communications* 86 (2016), Feb, Nr. 4, S. 1845–1856. http://dx.doi.org/10.1007/s11277-015-3160-1. – DOI 10.1007/s11277–015–3160–1

[LLM+16] LIU, Zishan ; LIU, Zhenyu ; MENG, Zhen ; YANG, Xinyang ; PU, Lin ; ZHANG, Lin: Implementation and performance measurement of a V2X communication system for vehicle and pedestrian safety. In: *International Journal and of Distributed and Sensor Networks* 12 (2016)

[LPM+15] LIU, Zhenyu ; PU, Lin ; MENG, Zhen ; YANG, Xinyang ; ZHU, Konglin ; ZHANG, Lin: POFS: A Novel Pedestrian-oriented Forewarning System for Vulnerable Pedestrian Safety. In: *International Conference on Connected Vehicles and Expo (ICCVE)*, IEEE, 2015

[LSH+18] LI, Chi-Yu ; SALINAS, Giovanni ; HUANG, Po-Hao ; TU, Guan-Hua ; HSU, Guo-Huang ; HSIEH, Tien-Yuan: V2PSense: Enabling

Cellular-based V2P Collision Warning Service Through Mobile Sensing. In: *IEEE International Conference on Communications (ICC)*, IEEE, 2018

[MBM11] Maslekar, N. ; Boussedjra, M. ; Mouzna, J.: A Stable Clustering Algorithm for Efficiency Applications in VANETs. In: *IEEE 7th International Wireless Communications and Mobile Computing Conference*, IEEE, 2011

[MHM+13] Morris, A.P. ; Hancox, G ; Martin, O ; Bell, D ; Johansson, C ; Rosander, P ; Scholliers, J ; Silla, A: Critical accident scenarios for cyclists and how they can be addressed through ITS solutions. In: *International Cycling Safety Conference*, 2013

[MLGN17] Math, Chetan B. ; Li, Hong ; Groot, Sonia H. ; Niemegeers, Ignas: A Combined Fair Decentralized Message-Rate and Data-Rate Congestion Control for V2V Communication. In: *IEEE Vehicular Networking Conference (VNC)*, 2017

[MOL15] Math, Chetan B. ; Ozgur, Ahmet ; Li, Sonia Heemstra H.: Data Rate Based Congestion Control in V2V Communication for Traffic Safety Applications. In: *IEEE Symposium on Communications and Vehicular Technology in the Benelux (SCVT)*, IEEE, 2015

[MSN17] Merdrignac, Pierre ; Shagdar, Oyunchimeg ; Nashashibi, Fawzi: Fusion of Perception and V2P Communication Systems for the Safety of Vulnerable Road Users. In: *IEEE Transactions on Intelligent Transportation system* Bd. 18, IEEE, 2017

[NAM17] Nasrallah, Yamen Y. ; AlAnbagi, Irfan ; Mouftah, Hussein T.: Enhanced Algorithms for the IEEE 802.11p Deployment in Vehicular Ad Hoc Networks. In: *IEEE 86th Vehicular Technology Conference (VTC Fall)*, IEEE, 2017

[NMKF20] Nguyen, Quang-Huy ; Morold, Michel ; KlausDavid ; Falko-Dressler: Car-to-Pedestrian communication with MEC-support for adaptive safety of Vulnerable Road Users. In: *Computer Communications* 150 (2020), S. 83–93

[NND12] Nagai, Makoto ; Nakaoka, Ken ; Doi, Yoshiharu: Pedestrian-to-vehicle Communication Access Method and Field Test Results.

In: *International Symposium on Antennas and Propagation (ISAP)*, IEEE, 2012

[Nsn20] NSNAM: *Network Simulator-3*. `https://www.nsnam.org/`. Version: 2020

[NYC15] NYCDOT: *NYC Connected Vehicle Project*. `https://www.cvp.nyc`. Version: 2015

[NYF+10] NAKANISHI, Yuki ; YAMAGUCHI, Ryohta ; FUJIMOTO, Kazushi ; WADA, Tomotaka ; OKADA, Hiromi: A New Collision Judgment Algorithm for Pedestrian-Vehicular Collision Avoidance Support System (P-VCASS) in Advanced ITS. In: *6th International Conference on Networking and Services*, IEEE, 2010

[OBZ17] OUNI, Sofiane ; BOULILA, Narjes ; ZAFAR, Bassam A.: Enhanced EDCA with Deterministic Transmission Collision Resolution for Real Time Communication in Vehicular Ad Hoc Networks. In: *Wireless Personal Communications* 98 (2017), aug, Nr. 1, S. 311–335. `http://dx.doi.org/10.1007/s11277-017-4871-2`. – DOI 10.1007/s11277–017–4871–2

[omn] *OMNeT++ Network Simulation Framework*. `http://www.omnetpp.org/`

[Per17] PERRY, Frank: *Overview of DSRC Messages and Performance Requirements*. `https://www.transportation.institute.ufl.edu/wp-content/uploads/2017/04/HNTB-SAE-Standards.pdf`. Version: 2017

[PIB11] PAUL, Bijan ; IBRAHIM, Md. ; BIKAS, Md. Abu N.: VANET Routing Protocols: Pros and Cons. In: *International Journal of Computer Applications* 20 (2011), Nr. 3, S. 28–34

[Pro17] *Prospect Project, Proactive Safety for Pedestrian and Cyclists*. Prospect Project,`http://www.prospect-project.eu/`, 2017

[QSW+18] QI, Weijing ; SONG, Qingyang ; WANG, Xiaojie ; GUO, Lei ; NING, Zhaolong: SDN Enabled Social Aware Clustering in 5G VANET Systems. In: *IEEE Access* 6 (2018), S. 28213–28224. `http://dx.doi.org/10.1109/access.2018.2837870`. – DOI 10.1109/access.2018.2837870

[RCB+16] Rostami, Ali ; Cheng, Bin ; Bansal, Gaurav ; Sjoberg, Katrin ; Gruteser, Marco ; Kenney, John B.: Stability Challenges and Enhancements for Vehicular Channel Congestion Control Approaches. In: *IEEE Transactions on Intelligent Transportation Systems* (2016). http://dx.doi.org/10.1109/tits.2016.2531048. – DOI 10.1109/tits.2016.2531048. – ISSN 1558–0016

[RCL+16] Rostami, Ali ; Cheng, Bin ; Lu, Hongsheng ; Kenney, John B. ; Gruteser, Marco: Performance and channel load evaluation for contextual pedestrian-to-vehicle transmissions. In: *1st ACM International Workshop on Smart, Autonomous, and Connected Vehicular Systems and Services (CarSys)* (2016). http://dx.doi.org/10.1145/2980100.2980103. – DOI 10.1145/2980100.2980103. ISBN http://id.crossref.org/isbn/9781450342506

[RKG18] Rostami, Ali ; Krishnan, Hariharan ; Gruteser, Marco: *V2V Safety Communication Scalability Based on the SAE J2945/1 Standard.* http://winlab.rutgers.edu/~rostami/files/pdf/rostami2018v2v.pdf. Version: 2018

[SCB12] Stanica, Razvan ; Chaput, Emmanuel ; Beylot, Andre-Luc: Congestion Control in CSMA-based Vehicular Networks: Do Not Forget the Carrier Sensing. In: *9th Annual IEEE Communications Society Conference on Sensor, Mesh and Ad Hoc Communications and Networks (SECON)*, IEEE, 2012

[SCZ+13] Shen, Xia ; Cheng, Xiang ; Zhang, Rongqing ; Jiao, Bingli ; Yang, Yang: Distributed Congestion Control Approaches for the IEEE 802.11p Vehicular Networks. In: *IEEE Intelligent Transportation Systems Magazine* 5 (2013), S. 50 – 61

[SGD11] Sommer, Christoph ; German, Reinhard ; Dressler, Falko: Bidirectionally Coupled Network and Road Traffic Simulation for Improved IVC Analysis. In: *IEEE Transactions on Mobile Computing* 10 (2011), S. 3–15

[SKCB12] Schaffer, B. ; Kalverkamp, G. ; Chaabane, M. ; Biebl, E. M.: A cooperative transponder system for improved traffic safety, localizing road users in the 5 GHz band. In: *Advances in Radio Science* 10 (2012), S. 39–44

[SKS17] SEWALKAR, Parag ; KRUG, Silvia ; SEITZ, Jochen: Towards 802.11p-based Vehicle-to-Pedestrian Communication for Crash Prevention Systems. In: *9th International Congress on Ultra Modern Telecommunications and Control Systems and Workshops (ICUMT)*, IEEE, 2017

[SMS+11] SEPULCRE, Miguel ; MITTAG, Jens ; SANTI, Paolo ; HARTENSTEIN, Hannes ; GOZALVEZ, Javier: Congestion and Awareness Control in Cooperative Vehicular Systems. In: *Proceedings of the IEEE* 99 (2011), jul, Nr. 7, S. 1260–1279. http://dx.doi.org/10.1109/jproc.2011.2116751. – DOI 10.1109/jproc.2011.2116751

[SNG10] SOUZA, Evandro ; NIKOLAIDIS, Ioanis ; GBURZYNSKI, Pawel: A New and Aggregate Local and Mobility (ALM) and Clustering Algorithm and for VANETs. In: *IEEE International Conference on Communications*, IEEE, 2010

[SNH08] SUGIMOTO, Chika ; NAKAMURA, Yasuhisa ; HASHIMOTO, Takuya: Prototype of Pedestrian-to-Vehicle Communication System for the Prevention of Pedestrian Accidents using both 3G and WLAN Communication. In: *3rd International Symposium on Wireless Pervasive Computing* IEEE, 2008

[SS19a] SEWALKAR, Parag ; SEITZ, Jochen: *Design considerations for the Cooperative ITS Vulnerable Road Users Service.* https://docbox.etsi.org/Workshop/2019/201903_STF565_VRU/A1-%20ETSI_STF565_VRUServices_TUIlmenau.pdf. Version: 2019

[SS19b] SEWALKAR, Parag ; SEITZ, Jochen: Vehicle-to-Pedestrian Communication for Vulnerable Road Users: Survey, Design Considerations, and Challenges. In: *MDPI Sensors: Enhances in V2X Communications for Connected Autonomous Vehicles* 19 (2019), S. 358

[SS20] SEWALKAR, Parag ; SEITZ, Jochen: MC-COCO4V2P: Multi-channel and Clustering based and Congestion Control and for Vehicle-to-Pedestrian. In: *IEEE Transactions on Intelligent Vehicles* (2020)

[SSS20] SHALA, Mendrit ; SEWALKAR, Parag ; SEITZ, Jochen: On-demand Quality-of-Service for Crucial Vehicle-to-Pedestrian Communication.

In: *16th International Conference on Wireless and Mobile Computing, Networking and Communications (WiMob)*, IEEE, 2020. – Submitted

[SYN+16] Swanson, Elizabeth D. ; Yanagisawa, Mikio ; Najm, Wassim ; Foderaro, Frank ; Azeredo, Philip: Crash Avoidance Needs and Countermeasure Profiles for Safety Applications Based on Light-Vehicle-to-Pedestrian Communications / National Highway Traffic Safety Administration. 2016. – Forschungsbericht

[TAA18] TAAS: *Ford starts testing V2X on public roads in China*. https://taas.news/article/105336/Ford_Starts_Testing_V2X_On_Public_Roads_In_China. Version: 2018

[TBL10] Taleb, T. ; Benslimane, A. ; Letaief, K. B.: Toward an Effective Risk-Conscious and Collaborative Vehicular Collision Avoidance System. In: *IEEE Transactions on Vehicular Technology* 59 (2010), mar, Nr. 3, S. 1474–1486. http://dx.doi.org/10.1109/tvt.2010.2040639. – DOI 10.1109/tvt.2010.2040639

[TIM17] TIMON: *Enhanced real time services for an optimised multimodal mobility relying on cooperative networks and open data*. https://www.timon-project.eu. https://www.timon-project.eu. Version: 2017

[TJHD13] Tielert, Tessa ; Jiang, Daniel ; Hartenstein, Hannes ; Delgrossi, Luca: Joint Power Rate Congestion Control Optimizing Packet Reception in Vehicle Safety Communications. In: *Proceeding of the tenth ACM international workshop on Vehicular inter-networking, systems, and applications*, ACM, 2013

[TLH+12] Thielen, Daniel ; Lorenz, Tobias ; Hannibal, Marco ; Koster, Frank ; Plaettner, Jens: A Feasibility Study on a Cooperative Safety Application for Cyclists crossing Intersections. In: *15th International IEEE Conference on Intelligent Transportation Systems*, IEEE, 2012

[TM13] Tal, Irina ; Muntean, Gabriel-Miro: V2X Communication-based Power Saving Strategy for Electric Bicycles. In: *Globecom Workshop - Vehicular Network Evolution*, IEEE, 2013

[TMGS13] TUNG, Lung-Chih ; MENA, Jorge ; GERLA, Mario ; SOMMER, Christoph: A Cluster Based Architecture for Intersection and Collision Avoidance Using Heterogeneous and Networks. In: *12th Annual Mediterranean Ad Hoc Networking Workshop (MED-HOC-NET)*, IEEE, 2013

[Tod19] TODAY, Traffic T.: *Japanese ITS waveband-based V2X system demonstrated in India*. `https://tinyurl.com/k5sbhjnd`. Version: 2019

[TP16a] TAHERKHANI, Nasrin ; PIERRE, Samuel: Centralized and Localized Data Congestion Control Strategy for Vehicular Ad Hoc Networks Using a Machine Learning Clustering Algorithm. In: *IEEE* 17 (2016), Nr. 11, S. 3275 – 3285

[TP16b] TAHERKHANI, Nasrin ; PIERRE, Samuel: Prioritizing and scheduling messages for congestion control in vehicular ad hoc networks. In: *ELSEVIER Computer Networks* 108 (2016), S. 15–28

[TSKF+15] TAHMASBI-SARVESTANI, Amin ; KAZEMI, Hadi ; FALLAH, Yaser P. ; NASERIAN, Mohammad ; LEWIS, Allan: System Architecture for Cooperative Vehicle-Pedestrian Safety Applications Using DSRC Communication. In: *SAE Technical Paper Series* (2015), Apr. `http://dx.doi.org/10.4271/2015-01-0290`. – DOI 10.4271/2015–01–0290

[USD14] USDOT: *Connected Vehicles: Vehicle-to-Pedestrian Communications*. `http://www.its.dot.gov/factsheets/pdf/CV_V2Pcomms.pdf`. Version: 2014

[USD18] USDOT: *Tampa (THEA) Pilot*. `https://www.its.dot.gov/pilots/pilots_thea.htm`. Version: 2018

[VRU16] VRUITS: *Improving Safety and Mobility of Vulnerable Road Users through ITS Applications*. `http://www.vruits.eu/`. Version: 2016

[WJS17] WILLIS, Jordan T. ; JAEKEL, Arunita ; SAINI, Ikjot: Decentralized Congestion Control Algorithm for Vehicular Networks Using Oscillating Transmission Power. In: *2017 Wireless Telecommunications Symposium (WTS)*, IEEE, 2017

[WLZF11] Wang, Qing ; Leng, Supeng ; Zhang, Yan ; Fu, Huirong: A QoS Supported Multi-channel MAC for Vehicular Ad Hoc Networks. In: *IEEE 73rd Vehicular Technology Conference (VTC Spring)*, IEEE, 2011

[WMY+14] Wu, Xinzhou ; Miucic, Radovan ; Yang, Sichao ; Al-Stouhi, Samir ; Misener, James ; Bai, Sue ; Chan, Wai hoi: Cars Talk to Phones: A DSRC Based Vehicle-Pedestrian Safety System. In: *80th Vehicular Technology Conference (VTC)* IEEE, 2014

[WSG+13] Wu, Xinzhou ; Subramanian, S. ; Guha, R. ; White, R. G. ; Li, Junyi ; Lu, K. W. ; Bucceri, A. ; Zhang, Tao: Vehicular Communications Using DSRC: Challenges, Enhancements, and Evolution. In: *IEEE Journal on Selected Areas in Communications* (2013)

[XCY18] XCYCLE: *XCYCLE Project*. `http://www.xcycle-h2020.eu`. Version: 2018

[YTH14] Yang, Fan ; Tang, Yuliang ; Huang, Lianfen: A Multi-channel Cooperative Clustering-Based MAC protocol for VANETs. In: *2014 Wireless Telecommunications Symposium*, IEEE, 2014

[ZGE18] Zadeh, Roya B. ; Ghatee, Mehdi ; Eftekhari, Hamid R.: Three-Phases Smartphone-Based Warning System to Protect Vulnerable Road Users Under Fuzzy Conditions. In: *IEEE Transactions On Intelligent Transportation Systems* 19 (2018), July, Nr. 7, S. 2086 – 2098

[ZGZ+19] Zhang, Degan ; Ge, Hui ; Zhang, Ting ; Cui, Yu-Ya ; Liu, Xiaohuan ; Mao, Guoqiang: New Multi-Hop Clustering Algorithm for Vehicular Ad Hoc Networks. In: *IEEE Transactions on Intelligent Transportation Systems* 20 (2019), Nr. 4, S. 1517–1530. `http://dx.doi.org/10.1109/tits.2018.2853165`. – DOI 10.1109/tits.2018.2853165

[ZSC06] Zhang, Xi ; Su, Hang ; Chen, Hsiao-Hwa: Cluster-based Multi-Channel Communications Protocols in Vehicle Ad hoc Networks. In: *IEEE Wireless Communications* 13 (2006), S. 44–51

[ZTL18] Zhang, Haibin ; Tian, Wanqing ; Liu, Jiajia: Improving EDCA for Efficient Channel Access in Vehicular Communications. In:

IEEE Communications Magazine 56 (2018), oct, Nr. 10, S. 72–77. `http://dx.doi.org/10.1109/mcom.2018.1800121`. – DOI 10.1109/mcom.2018.1800121

List of Figures

1.1 VRU fatalities by type and country 2
1.2 Dashcam footage of a self-driving vehicle just before fatal crash [Gua18] 3

2.1 Vehicle-to-Everything communication illustration 10
2.2 DSRC WAVE architecture [IEE19] 13
2.3 802.11p channel allocation for DSRC 14
2.4 Channel synchronization and intervals [IEE16a] 15
2.5 Channel access: (A) Continuous, (B) Alternating, and (C) Immediate [IEE16a] 15
2.6 WAVE MAC illustration 17
2.7 WAVE Networking Services 18
2.8 WSM packet format 22
2.9 WSMP-N-Header format 22
2.10 WSMP-T-Header format 22
2.11 WAVE Service Advertisement format 23
2.12 WSA header - mandatory fields 23
2.13 WSA header option indicator 23
2.14 WAVE Information Element Extension 23
2.15 WAVE Information Element 24
2.16 C-ITS station reference architecture [ETS10] 27
2.17 ITS-G5 access layer [ETS19] 27
2.18 ITS-G5 channel allocation 28
2.19 C-ITS Facilities layer functionality 31

3.1 Contact duration scenarios 43
3.2 Example V2P system architecture [SS19b] 46
3.3 Evaluation scenario 57
3.4 Channel Busy Percentage [SKS17] 60
3.5 CBP Distribution [SKS17] 61
3.6 Beacon packet Error Ratio [SKS17] 61

3.7 B_cDR for Pedestrian crossing in front of vehicle [SKS17] 63
3.8 B_cDR for Pedestrian walking parallel to direction of vehicle [SKS17] 63

4.1 Various pre-crash scenarios: **(a)** Pedestrian crossing in front of the vehicle, **(b)** Pedestrian moving parallel to the vehicle, **(c)** Vehicle turning left into the cyclist's path, **(d)** Vehicle turning right into the cyclist's path [SS19b] 69
4.2 Evaluation scenario [SS19b] 78

5.1 Clustering of pedestrians [SS20] 85
5.2 Evaluation scenario [SS20] 92
5.3 Channel Busy Percentage [SS20] 96
5.4 Total number of pedestrian beacons [SS20] 97
5.5 Average power consumption of pedestrian devices [SS20] 97
5.6 Maximum power consumption of pedestrian devices [SS20] 98

6.1 WSA Header Option Indicator [SSS20] 103
6.2 WAVE Information Element [SSS20] 104
6.3 Scenario [SSS20] 106
6.4 Crucial beacon delivery ratio for 25% node agreement [SSS20] . . . 109
6.5 Crucial beacon delivery ratio for 50% node agreement [SSS20] . . . 110
6.6 Crucial beacon delivery ratio for 75% node agreement [SSS20] . . . 110
6.7 Improvement in Crucial beacon delivery ratio [SSS20] 111
6.8 Channel Busy Percentage [SSS20] 111

List of Tables

2.1 Access categories in EDCA [IEE16d]. 16
2.2 EDCA QoS parameters [IEE16d]. 17
2.3 PSID Allocations [IEE16c]. 24
2.4 ITS-G5 Frequency Allocations [ETS13a]. 28
2.5 ITS-G5 channel grouping [ETS13a]. 28
2.6 Selected BTP port numbers [ETS16]. 29

3.1 Summary of current efforts for V2P systems 48
3.2 Simulation parameters [SKS17]. 59

4.1 Simulation parameters [SS19b]. 79
4.2 Scenario-specific parameters [SS19b] 79
4.3 Results [SS19b] 80

5.1 Cluster Notations 86
5.2 Cluster Parameters [SS20] 93
5.3 Simulation parameters [SS20] 94

6.1 Simulation parameters [SSS20] 107

Abbreviations

ABS .... Anti-lock Breaking System
AC .... Access Category
ACI .... Access Category Index
ADAS .... Advanced Driver Assistance Systems
AIFSN .... Arbitration Interframe Space Number
ART .... Available Response Time
BPSK .... Binary Phase Shift Keying
BSM .... Basic Safety Message
BSS .... Basic Service Set
C-ITS .... Cooperative ITS
C-V2X Cellular-V2X
CAM .... Cooperative Awareness Message
CBP .... Channel Busy Percentage
CBR .... Channel Busy Ratio
CCH .... Control Channel
CDMA .... Code Division Multiple Access
CH .... Cluster Head
CM .... Cluster Member
CSM .... Cluster-aware Safety Message
CSMA/CA .... Carrier Sense Multiple Access/Collision Avoidance
CW .... Contention Window
DCC .... Decentralized Congestion Control
DENM .... Decentralized Environmental Notification Message
DSRC .... Dedicated Short Range Communication
EDCA .... Enhanced Distributed Channel Access
ESC .... Electronic Stability Control
ETSI .... European Telecommunications Standards Institute
FCC .... Federal Communication Commission
FOV .... Field-of-View
GLOSA .... Green Light Optimal Speed Advisory

GPS	Global Positioning System
IEEE	Institute of Electrical and Electronics Engineers
IMU	Inertial Motion Sensor
IPv6	Internet Protocol version 6
ITS	Intelligent Transportation System
LIDAR	Light Detection and Ranging
LLC	Logical Link Control
LOS	Line-of-Sight
MAC	Medium Access Control
MC-COCO4V2P	Multi-channel Clustering-based Congestion Control for Vehicle-to-Pedestrian
MTW	Motorized Two-Wheeler
NLOS	Non Line-of-Sight
OBU	On-Board Unit
OFDM	Orthogonal Frequency Division Multiplexing
PDU	Protocol Data Unit
PRM	Priority Request Message
PSID	Provider Service Identifier
QAM	Quadrature Amplitude Modulation
QoS	Quality-of-Service
QPSK	Quadrature Phase Shift Keying
RADAR	RAdio Detection And Ranging
RSU	Road-Side Unit
SAE	Society of Automotive Engineers
SCH	Service Channel
STF	Specialist Task Force
SUMO	Simulation of Urban Mobility
TCP	Transmission Control Protocol
TDMA	Time Division Multiple Access
TPC	Transmission Power Control
UDP	User Datagram Protocol
UTC	Coordinated Universal Time
V2I	Vehicle-to-Infrastructure
V2N	Vehicle-to-Network
V2P	Vehicle-to-Pedestrian
V2V	Vehicle-to-Vehicle
V2X	Vehicle-to-Everything

VANET .... Vehicular Ad-hoc Network
VRU .... Vulnerable Road User
WAVE .... Wireless Access in Vehicular Environment
WME .... WAVE Management Entity
WSA .... WAVE Service Advertisement
WSM .... WAVE Short Message
WSMP .... Wave Short Message Protocol